18th Century Weapons of the Royal Welsh Fuziliers from Flixton Hall

3-16-02

To George-
With very best wishes
and many thanks for so much
help!

by Erik Goldstein

Thomas Publications
Gettysburg, Pa.

Printed and bound in the United States of America

Published by THOMAS PUBLICATIONS
P.O. Box 3031
Gettysburg, Pa. 17325

ISBN-1-57747-081-8

Cover design by Ryan C. Stouch

"To the brotherhood
of Royal Welch Fusiliers,
past, present & future"

Contents

Foreword

It was with a particular sense of pleasure that I accepted Erik Goldstein's request to write some opening remarks on his splendid piece of detective work relating to an object of great interest to us both: a group of British 18th Century military arms. To my knowledge this is the first time that such an investigation has been devoted to arms of non-royal or otherwise high profile origins or ownership.

The provenancing of antique objects goes back at least to the 17th Century, and was until the mid-20th Century largely confined to paintings and statuary, and those generally by the Great Masters and/or from Great Collections. After the end of World War II, with the departure of a great many art treasures, and some treasures not particularly artistic, in the hands of the victorious Allies (and others) from their long-time habitats, interest began to grow in where an item came from and where it has been since its creation. In some cases the fact that it came off the walls of a particular castle added in some degree to its worth, financial and sentimental. The down-turn in the fortunes of many families with ancestral collections of a wide variety of household goods and decorative features brought many items onto the antique market which had not changed hands for many decades, and it was found that the acquisitive urge was enhanced and encouraged if there was some background accompanying the object. Auction houses were quick to perceive the value of this enhancement and its application to an ever widening variety of objects proliferated. Unfortunately there was also a negative reaction to this new interest, resulting in the removal of many a crest, coat of arms or inscription from objects which the families selling them did not wish to be identified with their "degradation." This served to salve to some extent their pride, but also produced, in too many instances, lower prices than would otherwise have been realized: it has also been a great hindrance to scholars seeking to establish the precise dating or historical whereabouts of various objects.

The chief obstacle to accurate provenancing is vagueness and inaccuracy in the primary sources. In general, people making lists knew the objects they were listing and often failed to say anything in the notation which would distinguish the particular object from others like it, either then or later. This problem pervades the spectrum of objects being described, from royal and noble collections through military supply lists to probate inventories. Ignorant, semi-illiterate or simply unmotivated personnel making the lists were often unaware of significant distinctions, mis-spelt or misdescribed an object, or described it in a way as to leave open the precise identity of the particular object in question. Errors of omission as well as commission were also made, and it is sometimes significant why such omission might have been made.

One of the more recent arrivals on the scene of historical provenancing is antique arms, especially those of either artistic, mechanical and/or historic significance. As with more traditional objects, such efforts are usually directed towards pieces with royal or aristocratic associations or whose design and/or workmanship sets them above the general run of similar products. Antique military arms, particularly those bearing some form of individual issue markings associating them with a specific unit, are amongst the most recent objects to be subjected to the process of provenancing, although it may be said in a specific sense to have commenced some years prior to the Centennial celebrations of the United States in 1876. More generally, for some years there has been a strenuous effort to attribute a weapon with specific markings to a single individual who was present at a certain battle. Such efforts are usually unsuccessful, since the necessary military records do not often survive which would assign that weapon to an individual at any given date. The production of family records (dated diaries, correspondence, etc.) relating to a weapon of the same general type are normally far too vague to make the tremendous leap from document to object at all convincing, and yet wishful thinking has done much to span this yawning chasm.

It is, therefore, with a great feeling of relief that one can turn to this work of Erik Goldstein's and encounter scholarship which reduces wishful thinking to its barest minimum, and establishes lines of research along which one can travel with an increasing feeling of confidence that here, at last, the origins of a group of ordinary military products can be established with an acceptable degree of certainty. Having myself visited the ruins of Flixton Hall, and been able to furnish Erik with a number of pieces of primary source material, I thoroughly share both his interest and enthusiasm in this quest for background knowledge, especially since there is no denying that such knowledge, carefully and realistically confirmed, adds greatly to the appreciation of an historic object.

In an attempt to track down the origins and history of something as mundane and prosaic as a group of mid-18th century military muskets, bayonets, swords and polearms, Goldstein has broken new ground within the field of antiquarian provenancing, and as a student of history, I for one, apart from congratulating him, very much hope that he will start a trend as well as set standards for further serious enquiry in this important and neglected field.

De Witt Bailey
London, 2000

Author's Note

In the mid 1990s, when I mailed off the simple enquiry that ultimately lead to this book, the weapons about to be discussed had long been known to arms students on both sides of the ocean—but no one knew, or could remember, from where they came. Therefore, I suppose basic human nature had taken over, and the ensuing speculation led to an amusing array of "explanations" for the group's existence, a phenomenon that seems to have flourished for the past few decades. The stories I have heard range from the absurd; "they were captured from the Royal Welsh at Yorktown in 1781," or "they were made up for some English pageant in the 19th century," to the more realistic—the late Anthony D. Darling's explanation was very close to the truth. I had first become aware of the group in 1988, and after spending a good deal of time dreaming and wondering about it, decided to attempt to find the reality behind the lore. Based on the weapons involved, I knew it simply had to be extraordinary.

Dr. DeWitt Bailey, in the beginning of his foreword, most appropriately refers to this study as a "piece of detective work." As with a complex crime, through meticulous investigation one tries to "reconstruct" what has already happened—a very tricky feat for someone who did not witness the event or events in discussion. Consequently, no matter how sound the resulting story is, there will always be voids, some gaping and some inconsequential. Certain very important specifics concerning the collection of arms kept at Flixton Hall have proven unattainable—and will probably remain so.

First, most of the facts behind the formation of the collection are not likely to be discovered, since William Adair's papers as agent to the Royal Welsh Fuziliers have apparently not survived. Secondly, the disastrous fire at Flixton Hall in 1846 seems to have consumed a significant portion of the original assemblage. Thirdly, and most regrettably, no one thought to record the collection before its dispersal, leaving us to wonder—how many pieces, and of what types, were removed from the mansion in late 1950? What exactly did these displays of arms look like on the walls of Flixton Hall?

Surely, there are other weapons from this most remarkable hoard, in collections both public and private, which have yet to be identified as such. With this in mind, the author humbly requests to be contacted by anyone with information or pictures relating to any 18th century weapon marked to the Royal Welsh Fuziliers, William Adair or Flixton Hall.

Erik Goldstein
P.O. Box 1774
Williamsburg, Va. 23187
fuzilier@aol.com

Introduction

Caches, stashes or hoards, whichever term you prefer, are by nature, wonderful things. Wherever people have been, you find them. Most often, these terms describe some sort of "treasure," and are usually associated with great monetary value, such as a pot full of Roman gold found buried in an English farmer's field. At the other extreme, they can also be used to describe decades of old newspapers accumulated in the garage of an eccentric.

As applied to the field of antique weaponry, they are seldom encountered, and of these, only a tiny percentage has been brought to attention through publication. Those that have are most often from archaeological contexts, and are therefore somewhat incomplete or decomposed. The relative rarity of arms "hoards" is due to the fact that many of the weapons produced long ago, especially the mundane ones, were simply tools, and were subject to being "used up." This is doubly true for the weapons of the common soldier of centuries past.

With a great deal of luck, a "cache" of the common British foot soldier's arms of the 1720-1760 period managed to escape the hard use and eventual destruction for which they were specifically created. Had they not been almost immediately relegated to the wall displays of a gentleman's mansion tucked away in a quiet part of England, surely they would have fulfilled their destiny. Unfortunately, the magnificent 17th century manor house, which sheltered these arms for almost two centuries, fell victim to severe tax laws following World War Two, resulting in the "hoard" coming to light.

For the last half a century, this unique "cache" has all but completely escaped attention as an entity, in contrast to the lavish attention paid to many of its individual components. More than 15 years after this assemblage of weapons was dispersed, with the overwhelming bulk going to Colonial Williamsburg, they began to be noticed. The first specimens from Flixton Hall to be discussed in print, albeit not identified as such, appeared in George Neumann's 1967 edition of *History of the Weapons of the American Revolution*. Ever since, throngs of weapons students, authors, historians and re-enactors have looked to these items as the core of Colonial Williamsburg's collection of 18th century British military weaponry, now unequivocally known as the finest and most complete in the world. Without the weapons from Flixton Hall, there would be significant gaps in what we know about the subject. Taken as a whole, the weapons tell us far more about the arms themselves, the Board of Ordnance and the British Army of the mid-18th century, than any individual specimen ever has.

While the common British soldier's arms and accouterments of the 18th century have received quite a bit of attention in print over the last 30 or so years, these studies have focused primarily on their identification and typology. As a history of a group of weapons, the publication of the Adair Collection is unique, for a number of reasons. Most importantly, it is the sole group of military weaponry that can be traced back straight through one family to the famed regular-army unit to which they were originally issued. The group is additionally unparalleled because of the high state of preservation of most items. These factors, in the author's opinion, make the Adair Collection the undisputed "benchmark" for weapons of the types it includes.

As with any project of this scope, which draws from many different areas of study to give the most complete story possible, a great deal of help has been necessary. I have been fortunate enough to receive the most generous assistance of numerous institutions and individuals, many of which I have the honor of calling friends.

A tremendous debt of gratitude is owed to the following, for their absolutely invaluable contributions: Dr. DeWitt Bailey, Ted & Eileen Blackley, Lady Darell, Brigadier Sir Jeffrey Darell, Bt., Jay Gaynor, Curator of Mechanical Arts at Colonial Williamsburg, H. Russell Hall, Stuart Mowbray, Mr. David Parsons, Mrs. Ivy Parsons, and Mr. Stephen Wood.

In addition, I would also like to acknowledge the help of the following: William Ahearn, Diana Berry, Col. Peter Crocker, Curator, and David Bownes, Assistant Curator at the Royal Welch Fusiliers Museum at Caernarfon Castle, Leslie Carr, the late Anthony D. Darling, Glenn Earl, Susan Evans at the National Monuments Record, Christopher D. Fox, Curator of the Fort Ticonderoga Museum, Brian Godwin, Roger Gonzales, David Gray, Cathy Grosfils, Peter Hawkins, Robert Hendry, Geoffrey Jenkinson, Walter J. Karcheski, Jr., Senior Curator of Arms and Armor at the Frazier Arms Museum, Vincent J-R Kehoe, Albert Krassman, 446th Bomb Group, Fred Mahnken, 446th Bomb Group, Brian Malis, Kerry Meal, Searchroom Assistant at the Suffolk Records Office in Lowestoft, Clinton M. Miller, Stuart C. Mowbray, George C. Neumann, Graham Priest, the late Kit Ravenshear, William S. Rickman, 446th Bomb Group, Dr. John Ridgard, Dr. Arnold R. Saslow, Alison Strauss at the British Film Institute, Don Troiani, Dr. David S. Weaver, Timothy Wilson, and Thomas Wnuck.

Lastly, I must thank my wonderful wife Chelsea for her ubiquitous support, for enduring countless hours of my "Flixton-babble" and for braving the treacherous and unforgiving mud of East Anglia by my side.

Welch or Welsh?

Fusiliers, Fuziliers or Fuzileers?

Historians researching the early British army often encounter a variety of spellings for regimental names. For this particular unit, the question was settled in 1920 with the official adoption of the spelling ROYAL WELCH FUSILIERS.

However, throughout this study the author has chosen to use ROYAL WELSH FUZILIERS as the spelling of the name of the Regiment. Based on the examination of numerous period sources such as regimental barrel markings, officer's commissions and other manuscripts, this version seems to have been most prevalent during the period in question.

Since this study only relates to the Regiment during the relatively narrow span of circa 1740-1760, it seems most appropriate to use the spelling most common at the time.

William Adair (1700-1783) probably by Francis Hayman (1708-1776), oil on canvas. Executed c. 1760-70, the noted army agent is depicted in front of his recently purchased home, Flixton Hall. Note that by this time, the moat, clearly visible in the 1752 view (see page 22), has been filled in. Adair is wearing a dark blue suit with gold buttons, a gold-lace trimmed cocked hat, buff gloves, black riding boots, and is holding a riding crop. Another portrait of William Adair, with his groom and two horses, was destroyed in the fire at Flixton Hall in 1846 (ILN, 19 December 1846). Courtesy of Lady Darell.

William Adair, John Huske and the Royal Welsh Fuziliers

The Adair family, who are descended from the Fitzgeralds, were seated at modest Dunskey Castle at Portpatrick, in Wigtonshire, Scotland during the early 17th century. At the time, there was civil unrest in Ireland centered upon Ulster, and in an attempt to keep control of the situation, King James I encouraged English and Scottish citizens to settle there. In 1610, the King gave the family some land in Bally-mena, County Antrim, and a branch of the Adair family left Scotland to live there. By the early 18th century, the Adair family had established a seat at Ballymena Castle, and had helped introduce a prosperous linen industry to the area. Until very recently, land in Ballymena was still held by the Adairs.

The earliest member of the family to be given a title was Robert Adair, created a Knight Banneret for his conduct at the Battle of the Boyne in 1690. Robert had raised a regiment under William III, and was amongst the last to receive that title. His son and successor, William Robert Adair, was a Captain in Lord Mark Kerr's Regiment of Horse, and fought at the Battle of Culloden in 1746.

It is to such a distinguished family that the William Adair[1] who is the focus of this study was born in Kirkmaiden, Galloway, Scotland in 1700. He was the eldest child of a large family born to parents who were distantly related and part of the large network of Adairs living in that section of Scot-

Ballymena Castle, seat of the Adair family in County Antrim, N. Ireland, as seen in an early 20th century postcard. After a fire in the 1870s, Lord Waveney had it rebuilt by William Henry Lynn in the Scottish Baronial style. Unlived-in since WWI, it burned down again in 1956, and its site is now occupied by houses, a library and a swimming pool. Courtesy of Lady Darell.

1736 Copper twopence token issued by James Adair of Ballymena, who was probably connected with the family's linen industry. During the period, merchants often issued tokens for the benefit of local commerce. The obverse depicts a rabbit with the legend "I MAKE GOOD SPEED," and the reverse reads "I PROMISE TO PAY THE BEARER ONE (sic) DEMAND TWO PENCE IAS ADAIR B MENA 1736" (Davis, Antrim #2). Author's collection.

land and Ulster. Unfortunately, other than the fact that he was educated at Carrickfergus, Ireland, little else is known about his formative years. By the early 1740s, William Adair was quite actively employed as an army agent in London.

The function of the army agent was an essential one but to be understood, it must be examined in context, taking into account the way the British Army of the period was organized and managed.[2] From the late 17th century, a system of proprietary Colonelship had been in place, with units which somewhat resembled franchises. Regiments were raised by men of status, who were as a rule wealthy, and often substantial landowners. Many of these Colonels also held titles, and served in civic capacities as well. Parliament then voted monies to cover most of the expenses of the regiments, which was then issued to the Colonels for disbursement.

Therefore, the responsibility for the discipline, supply and finance of each regiment was solely in the hands of its Colonel, who may additionally have held General Officer's rank. Since he needed to give his attention to military matters, the Colonel would hire a civilian "agent" to handle his regiment's business matters, amongst other various professional and personal duties.

The position of agent was obviously a very lucrative one if carried out properly, since Colonels occasionally offered the position to the highest bidder. It can be assumed that since such a high price was paid, the return was expected to be in line with the investment. Additionally, an agent might have to put up money as security, for protection against his potential ineptness or corruption putting the regiment and its Colonel into debt. In 1746, one agent paid 850 Pounds Sterling to obtain the custom of a Dragoon regiment because his predecessor left a debit of 3000 Pounds Sterling, which the Colonel was forced to make good.[3]

Although post-dating the height of William Adair's career, a treatise on military finance was published for the use of army officers, paymasters and agents in 1782. The following was said about the pay of agents;

> ...the allowance consists of a man per Troop or Company and 2 pence in the Pound from the whole Pay of the Regiment; that the place of Agent is sometimes a gift but generally a purchase; that he is properly the Colonel's clerk and it is at the Colonel's option whether he shall receive the whole or any part of the agent's allowance except the 2 pence in the Pound, being obliged to account to the Colonel for the rest if he thinks fit to reserve it for his own use.[4]

In plain English, there were two allowances due to the agent. The first was fixed at 2 pence per Pound of the gross pay of the unit, which was deducted from the off-reckonings. The term "off reckonings" refers to the 25% of the soldier's daily pay held back by the Colonel to pay for expenses such as the soldier's uniforms and accouterments. Additionally, some of this money went towards the upkeep of Chelsea Hospital.

The second was the pay of the "agent's man" in each troop or company of the unit, per diem. In times of war, when the units were at their largest, the agent's pay could be on par with that of the regimental officers, depending on the agreement he had with the Colonel. In other cases, both of these allowances were waived by the agent in favor of a fixed stipend.[5] It is also certain that there was quite a bit of private business going on, such as money lending, which put additional money into the agent's pocket.[6]

Once an agent had been appointed by the Colonel, approval of the Secretary at War was required. Once received, it was then arranged for the agent to act, under power of attorney, for the Colonel (see page 21). The duties of the agent, as detailed by this power were;

> To ask demand receive of and from...the Paymaster-General for the time being, all such sum or sums of money as is now or may hereafter become due and payable unto me or the aforesaid Regiment under my command, as well as officers as soldiers thereof, either for pay, off-reckonings, arrears, allowances or any other account whatsoever; and also contract for the clothing, accoutrements, etc., of the Regiment...and upon receipt of any sum or sums of money for subsistence, pay...or any account whatsoever as aforesaid, receipts and proper discharges for me and in my name to give and to do and to execute all and every other matter act or lawful thing needful and necessary to be done in or about the premises, hereby ratifying, confirming and allowing all and whatsoever my said Agents and Attorney shall lawfully do or cause to be done by virtue of these presents."[7]

Adair House, in Pall Mall, as it appeared c. 1814. Built in the mid-18th century, this house stood on the northeast corner of George Street (with its rear facing St. James' Square) until 1886. It was from this building that William and Alexander Adair conducted their business as army agents. The family retained the house until the mid-19th century. Author's drawing, after one in Sheppard's Survey of London, Vol. XXX (in the plan pocket), based on an original sketch in the Crace Collection at the British Library. Author's copyright.

The agent was also accountable to the War Office for disbursements on account of his regiments. The Paymaster-General made payments to the agents, who in turn made payments to the regimental paymasters, who in turn made disbursements to the company Captains. These regimental paymasters were officers appointed by the Colonel to act as such, and still had their ordinary military duties to perform.

At this time, most of the supplies needed by the regiments were privately procured, and it was the agent who handled this too. He dealt with clothiers, sword-cutlers, gunmakers and insurance agents, amongst others. He also dealt with the Board of Ordnance concerning the arming of the regiment. When such items as muskets, bayonets and cartridge boxes were needed, the agent would petition the Ordnance for them, and act as the conduit between that office and the regiment. Although swords were still carried by infantry soldiers at this time, they were not issued to the regiments from Ordnance stores. Since in the infantry regiments, they were more ornamental than functional, they were classified as accouterments, and therefore were privately contracted-for by the agents. Additionally, agents acted for the Colonel and his regimental officers in numerous other private and official capacities, such as assisting in the sale and purchase of officers' commissions.

Like many of the other agents, William Adair set up offices in one of his London residences. Built c.1750, and known as "Adair House," this four story Georgian building stood on the northeast corner of Pall Mall and George Street, until it was torn down in 1886. Adair also employed Henry and Thomas Bullock as his clerks. These men also went on to become career army agents.[8]

In such a business, war meant prosperity. During peacetime, regiments were disbanded, and others were reduced. Fewer soldiers meant less money from the government, and less profit for the agents. Therefore, agents sought the employ of as many regimental Colonels as they could get, or thought they could handle. John Calcraft, by far the busiest of the agents of the era, managed 63 out of 160 regiments at the height of the Seven Years War.[9] William Adair could handle a much smaller share, and was recorded as the agent to the following 9 regiments in 1745;[10]

William Adair's Regiments, 1745

3rd Troop of Horse Guards (Tyrawley's)
2nd Horse (Honeywood's)
2nd "Coldstream" Guards (Earl of Albemarle's)
19th Foot (Howard's)
23rd Foot or Royal Welsh Fuziliers (Huske's)
32nd Foot (Douglas's)
33rd Foot (Johnson's)
44th Foot (Kennedy's)
6th Marines (Cotterell's)

However, the list of an agent's clients was constantly changing. Since they worked so closely with their Colonels, and often became personal friends, a change in a regiment's Colonel often meant a change in the agent too. By late 1761, of Adair's then 9 clients, only three had remained since 1745. They were:[11]

William Adair's Regiments, 1761

3rd Troop of Horse Guards (Tyrawley's)
3rd Dragoons
5th Foot
9th Foot
12th Foot
19th Foot (Green Howard's)
23rd Foot (Royal Welsh Fuziliers)
38th Foot
72nd Foot

One of the most fruitful of Adair's agencies was that with General John Huske. Adair had been Huske's agent while he was Colonel of the 32nd Regt. (from 1740), and kept the position as his agent when Huske got the Colonelcy of the 23rd Regt. in 1743 (he also managed to hang on to the agency of the 32nd Regt. for a while too).

John Huske is thought to have been born around 1692, and led an exceptional life in both the military and political arenas. He was first commissioned as an Ensign in Col. Toby Caulfield's Regiment of Foot in 1708, and from there obtained a Company in Lord Hertford's Regiment of Foot (15th) in 1715. By 1734, he was the second Major of the Coldstream Guards, and on Christmas Day 1740, received the Colonelcy of the 32nd Regiment of Foot. At the Battle of Dettingen in 1743, he was serving as a brigadier, and was reported to have "behaved gloriously," receiving severe wounds in the action.[12]

In recognition of his gallant services at Dettingen, he was promoted to Major General and was chosen to be the new Colonel of the Royal Welsh Fuziliers, in place of Colonel Newsham Peers, who had been killed at the same battle. During the Jacobite Rebellion of 1745-6, he was the second-in-command at the Battle of Falkirk, where he secured the retreat of the Royal forces to Linlithgow, thereby preventing a bad situation from getting much worse. At Culloden, he again served as second in command, but this time under the Duke of Cumberland. Afterwards, he was promoted to Lieutenant General, and served in Flanders during 1747-8. On 5 December 1756, he was promoted to General. In addition to his military responsibilities, he was appointed Governor of Sheerness in 1745 and Governor of Jersey in 1760. These were both positions which did not require residency, and served to act as extra income or a pension for Huske.

Colonel Newsham Peers of the Royal Welsh Fuziliers, by an unknown artist, c. 1743. Peers first entered the Regiment c. 1705, while his uncle, General Joseph Sabine, was its Colonel. His service to the Regiment was distinctive, and when his uncle died at Gibraltar in 1739, Peers succeeded him to the Colonelcy. Peers wears a red coat with gilt buttons set in pairs and royal blue cuffs, red breeches and a richly-embroidered gold waistcoat. He holds buff leather gauntlets and a gold-laced black cocked hat, and is wearing what appears to be a small sword with a hilt of gilt metal. Peers, the last Colonel to lead the Regiment in battle, was killed at the Battle of Dettingen (depicted in the left background) on 27 June 1743. Courtesy of the Trustees of the Royal Welch Fusiliers Museum.

General Huske was apparently an outstanding soldier, later described as a "brave, blunt veteran."[13] Regardless of his duties outside and above his regiment, he saw that it maintained its status as one of Britain's finest infantry regiments. The concern he showed for his soldiers even earned him the affectionate nickname of "Daddy Huske."

A military man of Huske's caliber was perfectly suited for a regiment as esteemed as the Royal Welsh Fuziliers.[14] The regiment had been formed in the Welsh marshes near Ludlow in April of 1689 for William III's army fighting against that of the deposed James II in Ireland, and saw its first action at the Boyne in 1690. During the War of the Spanish Succession, the Regiment served under the Duke of Marlborough, and was first designated as a "fusilier" regiment in 1704. As Fuziliers, the Regiment's function was to protect the artillery, so they were armed with flintlock muskets, or "fusils," instead of the still-common matchlock muskets. This was done out of concern for safety, since matchlock muskets required the use of burning match, an obvious hazard in the presence of the large quantities of the artillery's gunpowder. Battle Honours earned while serving under Marlborough include those for the Battles of Blenheim, Ramillies, Oudenarde and Malplaquet. "Royal" became part of the Regiment's title in 1714, in recognition of their service during the war.

After Dettingen (1743), famous as the last battle in which a British King (George II) led his troops, the Regiment became Huske's. The Royal Welsh saw action on the Continent in the 1740s, fighting at Ghent, Fontenoy and Laffeldt. By then numbered the 23rd Foot (seniority numbers were given in 1751), the Regiment was part of the garrison of Fort St. Philip at Port Mahon in Minorca when the French successfully captured the island in 1756.

Probably the most prized Battle Honour awarded to the Royal Welsh was for the Battle of Minden, fought on 1 August 1759, and celebrated by the Regiment on every anniversary since. As part of the army of Prince Ferdinand of Brunswick, the Regiment was one of four British infantry units which helped draw the French forces from their cozy garrison. In the melee of battle, as a result of misunderstood orders, the British infantry marched steadily into the middle of the field with beating drums. The actual order had been "to attack at the beat of the drum." When the British infantry was quickly attacked by the French cavalry, all were shocked to see the French repulsed three times, with murderous results. This unparalleled infantry action broke the momentum and morale of the French army, and was instrumental in winning the day for the allies. Unfortunately, Huske was not with his Regiment on that glorious day.

The relationship between Adair and Huske obviously went far beyond that of a solely professional one, and can be accurately be described as a close friendship. When Huske died in January 1761, William Adair was the residual legatee of his estate, in addition to being appointed his executor.[15] Adair's inheritance from Huske was valued at 16,000 Pounds Sterling, a huge sum for the day. This inheritance also included the General's estate outside in London called Ealing House, and any personal property in it.[16]

Although William Adair's agency with the Royal Welsh Fuziliers ended with the death of General Huske, his business continued to flourish. Over the course of his career, he had served as agent to such distinguished military men as the Duke of Cumberland, General Studholme Hodgson, the Earl of Albemarle, the Duke of Richmond, Lord George Lennox and General Robert Napier. Adair also appealed to important naval heroes as well, serving as agents to Lord Anson and Admiral Lord Keppel.[17]

Grenadier Private of the Royal Welsh Fuziliers, from David Morier's painting of the Grenadiers of the 22nd, 23rd and 24th Regiments, c. 1751. This confident soldier wears a red coat faced with Royal blue and laced with the blue, yellow and red lace of the Regiment. His waistcoat is red, his breeches are Royal blue, and his parade gaiters are white with black buttons. He carries an iron-hilted hanger and his accouterments are of buff and black leather. Of special note is his mitre cap, embroidered with the three devices found on the grips of the hangers made for the Regiment. The Royal Collection, Copyright, Her Majesty Queen Elizabeth II.

22 *Will of Lieut. Gen.* Hufke.

Copy of the WILL *of the late Lieutenant General* HUSKE.

	£.	s.
TO Mr *Nichols* of *Ealing*	10	10
To 24 poor women of *Ealing*, 20 *s.* each	24	0
To Mr *William Chamberlayne*, his attorney	52	10
To Mr *Treufdale*, his apothecary	100	0
To the poor of *Newmarket*	100	0
To Lord *Cadogan* ten pictures.		
To the Earls of *Godolphin* and *Winchelfea*, Lord *Berkeley* of *Stratton*, Col. *Southeby*, Major *Caulfield*, *John Anthony Bellaguier*, the Ladies *Cadogan*, *Hyde*, *Margaret Bentinck* and *Codrington*, & Mrs *Finch*, 100 guineas each	1155	10
To *George Horton*, an upholder in *Piccadilly*	500	0
To the Earl of *Albemarle*, Ld *Hyde*, Major Gen. *Yorke*, and *Edward Finck*, Efq; 1000 *l.* each	4000	0
To the Earl of *Granville*	2000	0
To Capt. *D'Auvergne*	2000	0
Total of legacies to his acquaintance the poor, &c. is	9942	10
To *Thomas*, a labourer in his garden	10	0
To *Daniel*, his under-gardener	20	0
To *John Chipps*, his gardener	50	0
To *Sarah Booth*, *Eliz. Weetbley*, and *Catharine Kelly*, houfemaids, 100 *l.* each	300	0
To *John Thatcher*, a lad that looked after his cattle	120	0
To *Margaret Lloyd*, houfekeeper at *Ealing*	150	0
To *Wm Garftin* his poftillion	500	0
To *James Lawton*, his under groom	500	0
To *Mary Knights*, his town houfe keeper, 30 *l. per ann.* during her life, valued at 15 years purchafe, and 50 befides	500	0
To *Edw. Webb*, his footman	1200	0
To *Jofeph Mofum*, his valet de chambre, 200 *l. per ann.* during his life, valued at 15 years purchafe, and 50 *l.* befides	3050	0
Said valets daughter, when of age, or married	1000	0
Said valets wife, in cafe fhe furvives her hufband, 50 *l. per ann.* for her life		
To the faid valet and footman, all his cloaths, linen, and wearing apparel		
To *Thomas Calridge*, his groom of the ftables	5000	0
All his horfes, horfe furniture, and wheel carriages, are alfo given to the faid groom, the value of which not yet known		
A To one year's wages and three months board wages to all his fervants		
Total, of his bequefts to his fervants, exclufive of apparel, horfes, horfe furniture, carriages, & wages is	12400	0
B To the two daughters of his his late brother *Ellis Hufke*, 40 *l. per ann.* during their lives, valued at 15 years purchafe - - -	1200	0
C To the four grand-children of his late brother *Richard Hufke*, 500 *l.* each	2000	0
Total of his bequefts to his family, provided the legatees do not difturb the executors in the execution of the will, by any fuit in law or equity, is - - -	3200	0
D To Mr *Henry Bullock* 200 *l.* and Mr *Tho. Bullock*, 100 *l.* both clerks to *Wm Adair*, Efq; and joint executors with him	300	0
E To *Wm Adair*, Efq; agent to his regiment and government, and one of the executors, the houfe and land at *Ealing*, which coft 2800 *l.* but is now worth, on account of improvements, &c.	4000	0
F The faid agent is refiduary legatee, which is known to be worth (exclufive of the furniture and wine-cellars of town and country-houfe, and the falling in of all the above annuities)	12000	0
Total to his executors, is	16300	0
G	41842	10

Dialogue between Pafquin *and* Marforio; *faid to have been tranfmitted from* Rome.

Pafquin. WELL, *Marforio*, are not thefe fine doings in the North?

H *Marforio.* Yes, truely; thoufands cutting each others throats, they know not for what.

Paf. What! is not a province worth contending for?

Mar. No; the blood and treafure it has

(Left) The will of General John "Daddy" Huske from the January 1761 issue of The Gentleman's Magazine. *Huske's bequest to William Adair and his assistants, Henry and Thomas Bullock, are the last three entries. The fact that Huske so generously bequeathed money to his servants and dependants perfectly illustrates one of the reasons he earned his paternalistic nickname, and is possibly why his will was published in this forum. Author's collection.*

(Below) Ealing House, the home of General Huske, as it appeared in 1795 (watercolor, 22 x 17cm). At the time of Huske's death in 1761, this mansion and its land, on account of recent improvements, was valued at 4000 Pounds, and was bequeathed to William Adair. It was demolished in the 19th century. Guildhall Library, Corporation of London.

According to the Army Lists, some time before 1767,[18] William Adair took his nephew Alexander, who seems to have been more like a son, under his wing and brought him into the business. Additionally during the 1760s, Adair began partnering some of the agencies with Henry or Thomas Bullock, or possibly both. These two men were certainly related, but it is unclear how—they were likely either father and son or brothers.

By the era of the American Revolution the business had evolved, and William Adair, now in his 70s, probably handed over the reins to Alexander. Instead of a sole proprietorship in the name of William Adair, there was a confusing assortment of partnerships and sole proprietorships acting as agents. Some regiments listed Mr. Bullock as their agent, some listed Mr. A. Adair, some listed just Mr. Adair, and others listed Messrs. Adair and Bullock. To complicate the matter further, agencies and partnerships seem to have changed from year to year.

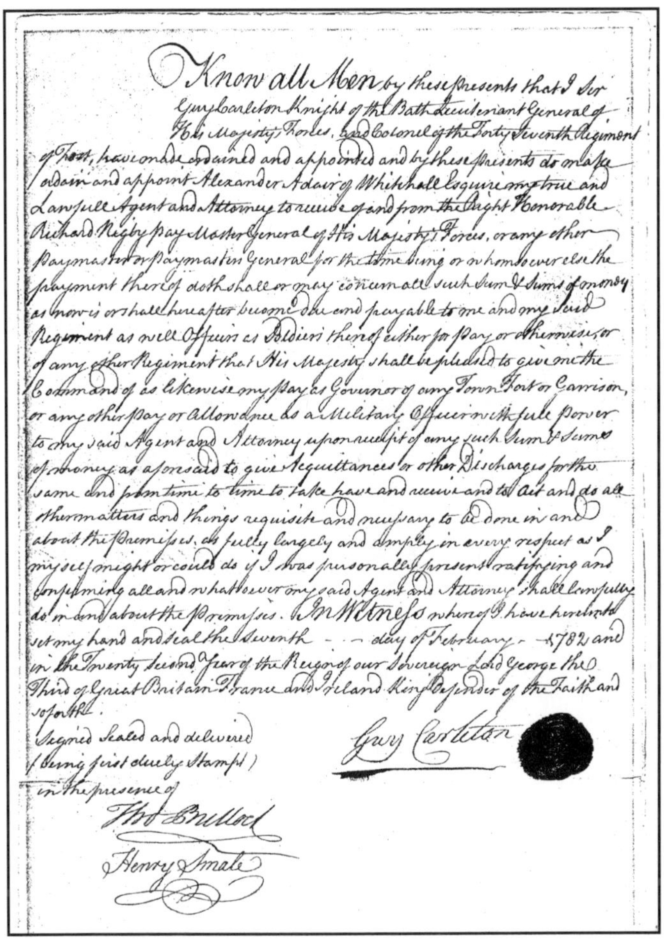

Know all Men by these Presents that I Sir Guy Carleton Knight of the Bath Lieutenant General of His Majesty's Forces, and Colonel of the Forty Seventh Regiment of Foot, have made ordained and appointed and by these presents do make ordain and appoint Alexander Adair of Whitehall Esquire my true and Lawfull Agent and Attorney to receive of and from the Right Honorable Richard Rigby Pay Master General of His Majesty's Forces, or any other Paymaster or Paymasters General for the time being or whomsoever else the payment thereof doth shall or may concern all such Sum & Sums of money as now is or shall hereafter become due and payable to me and my said Regiment as well Officers as Soldiers thereof either for Pay or otherwise or of any other Regiment that His Majesty shall be pleased to give me the Command of as likewise my Pay as Governor of any Town Fort or Garrison, or any other Pay or Allowance as a Military Officer with full Power to my said Agent and Attorney upon receipt of any such Sum & Sums of money as aforesaid to give Acquittances or other Discharges for the same and from time to time to take have and receive and to act and do all other matters and things requisite and necessary to be done in and about the Premises, as fully largely and amply in every respect as I myself might or could do if I was personally present ratifying and confirming all and whatsoever my said Agent and Attorney shall lawfully do in and about the Premises. In Witness whereof I have hereunto set my hand and seal the Seventh day of February 1782 and in the Twenty Second Year of the Reign of our Sovereign Lord George the Third of Great Britain France and Ireland King Defender of the Faith and so forth.

Signed Sealed and delivered (being first duly Stampt) in the presence of
Thos Bullock
Henry Smale

Guy Carleton

The original document officially appointing "Alexander Adair of Whitehall, Esquire" as agent to Lt. General Sir Guy Carleton and his regiment, the 47th Foot. At this time, Carleton was also serving as Commander in Chief of His Majesty's forces in America. Thomas Bullock and Henry Smale have signed as witnesses in lieu of Alexander Adair, and the reverse bears manuscript dockets indicating that it was entered in the offices of John Lord Viscount Mountstuart, Lewis Lord Sondes and Richard Rigby (as the late Paymaster General of His Majesty's Forces). Author's collection.

By 1781, "Messrs, Adair and Bullock, Whitehall" only had the agency of 2 cavalry and 2 foot regiments, in addition to that of eight militia regiments. These units, and their Colonels were:

Cavalry

1st Troop of Horse Guards
(Wm., Marquis of Lothian)
22nd Light Dragoons
(John Baker Holroyd)

Regiments of Foot

25th Regiment (Lord George Lennox)
47th Regiment (Sir Guy Carleton)

Militia

Cumberland (James Lowther)
Devonshire (Paul Orcherd)
S. Glocestershire
(Earl Frederick Berkeley)
E. Kent (John Sawbridge)
Pembroke (John Colby)
Shropshire (Lord Clive)
Sussex (Duke of Richmond)
Westmoreland (Sir James Lowther)

Alexander Adair, along with Thomas Bullock, carried on the business of army agency long after William Adair's retirement and eventual death. Sadly, although the careers of the two Adairs as agents spanned a more than 70-year period between the War of the Austrian Succession and the Napoleonic Wars, almost none of their military-related papers appear to have survived. However, there are a few in the Adair papers at the SRO, Lowestoft and a very small number in private hands.[19] A few letters written by William Adair also survive at the British Library amongst the papers of the Duke of Newcastle and Lady M. Bentinck.

Flixton Hall and the Adairs

Until its demolition in the 1950s, Flixton Hall had the reputation of being one of the finest houses in the County of Suffolk. While its exact date of construction is unknown, it was built circa 1615 for John Tasburgh on what was probably the site of the ancient manor house of the Bateman family. Little is known of this earlier structure, other than that it was once occupied by the William Bateman who was consecrated as Bishop of Norwich in 1343. By tradition, Flixton Hall is thought to have been the work of Inigo Jones (1573-1652), a very well known architect of the late Elizabethan and early Stuart era, but this has not been proven. It seems more likely that, if anything, he was responsible for remodeling it. Jones's association with Flixton Hall, whatever it was, was important enough for a bust of him to be mounted above the doorway to the salon on the second floor, which remained in place until it was destroyed in 1846.[20]

The Tasburghs had originated in Norfolk, and settled in the Flixton area in the 13th century. Being Catholic, the Tasburghs were suspected of being involved when the town of Bungay was all but destroyed by fire in 1688, but were spared persecution because of the generous relief they afforded to many of the panic-stricken residents of the town.

An amusing account of Flixton Hall in the 17th century pertains to King Charles II, who was passing in front of the house during one of his "progresses," probably on the old road connecting Homersfield and

A View of Flixton Hall from the West End of the Upper-Terrace, executed in watercolor in 1752 by Thomas Sandby (16 1/2 x 29 1/2). As the earliest locatable view of the mansion, it is the only one that shows the moat. This painting was later engraved in 1787 by W. Angus for his Seats of the Nobility.... *Courtesy of Lady Darell.*

The bust of Inigo Jones, the architect credited for Flixton Hall's design, as it appeared above the doorway to the salon on the second floor of Flixton Hall. This bust was destroyed in the fire of 1846. From the 26 December 1846 issue of the Illustrated London News. *Author's collection.*

Bungay. The King apparently admired Flixton Hall, and asked one of his attendants whose house it was. It belongs "to a Popish dog, Your Majesty," replied the attendant, to which the King responded "The dog has a very beautiful kennel."[21]

The Tasburghs remained at Flixton until the mid-18th century, when the direct male line became extinct and the estate passed into the Wyborne family, who only retained it for about fifteen years. In 1752, a member of the Wyborne family commissioned the noted artist Thomas Sandby to paint a watercolor of the great house. This is the earliest image of Flixton Hall yet discovered, and shows the house in what is probably close to its original state. While the once vivid colors of the painting have faded over the years, we can see that it was constructed of patterned red brick with off-white stone stringcourses and window dressings. The basic outline of the structure was shaped like an "E" (some say in honor of Queen Elizabeth I), and many of its corners were adorned with octagonal shaft-like buttresses. The flat roof was embattled, and it bristled with ornamental brickwork chimneys and pinnacles. Facing to the north, around the whole, was a moat traversed by an arched brickwork bridge which lead to the main entrance in the center of the building.

The next year, in 1753, the estate was sold to William Adair, the wealthy army agent. Adair apparently adored his home, the acquisition of which brought him "landed gentry" status with its many social and business perks. He spared no expense furnishing it appropriately, had his portrait painted in front of it, and arranged for his tombstone in Flixton Church to be inscribed "William Adair, Esq. of Flixton Hall" after his death.

Shortly after his purchase of Flixton Hall, he began to alter it significantly, in what was to be the first in a series of alterations made to the mansion by the Adair family over the following century and a half. The moat and bridge, clearly visible in the 1752 Sandby painting, had disappeared by the time William Adair had been portrayed in front of it.[22] Further confirmation of this appears in *The Seats of the Nobility and Gentry in Great Britain and Wales....* published by W. Angus in 1787. Plate VIII is a very accurately engraved version of Sandby's 1752 view, with a description, which notes that "The Terrace shewn (sic) in the View of this mansion has been leveled (sic), and the moat filled up, since the present Drawing was taken."

There were other pre-1787 alterations to Flixton Hall according to Angus—"To the South was an open Portico and Colonnade, now closed up, and converted into separate Rooms."

When William Adair died, in May 1783, he left a 13-page will[23] relating to his personal estate, and a separate will pertaining to his real estate. Like General Huske, William Adair bequeathed generous sums of money for charitable uses, benefiting the parishes in which he was born, educated and lived. He also left money to various family members and dependents. His clerk, and one of the co-executors of the will, Thomas Bullock, received 100 Pounds Sterling, as did his two other clerks, Mr. Oglivie and Mr. Doubleday.

However, since his only child was a daughter born out of wedlock,[24] he left the bulk of his wealth to his nephew, Alexander Adair. Included in Alexander's inheritance was the Flixton Hall estate with all of its furnishings and heirlooms (for Alexander's

Lord Anson medal, 1747, commemorating the defeat of the French fleet off Cape Finisterre, and his 1740-44 voyage around the world. William Adair was also the agent for Lord Anson, and cherished a solid gold striking of this medal highly enough to mention it specifically in his will. This specimen is in gilt brass (Hawkins, Medallic Illustrations ii, p.634, #325).

lifetime only) and the business as an army agent, in addition to Adair House in London as well.

Some items of interest specifically mentioned in William Adair's will were:

> A ring of the value of 100 pounds Sterling was bequeathed to both Lord George Lennox and Lt. General Hodgson.
>
> His large diamond ring was bequeathed to Augustus Lord Keppel (the Admiral).
>
> A gold-hilted sword presented to William Adair by the Earl of Albemarle, and a gold medal of Admiral Lord Anson, were to remain with his estate as heirlooms.

William Adair's daughter, Jane Smith Adair Brice, and her children, were well provided for, regardless of not being the main beneficiaries of the estate. Edward Brice, who had wed Jane in August of 1758,[25] was one of the co-executors of his father-in-law's estate.

The monument to William Adair erected by Alexander Adair in Flixton Church. The family arms appear in the pediment, and there are cow-skull supporters at the bottom.

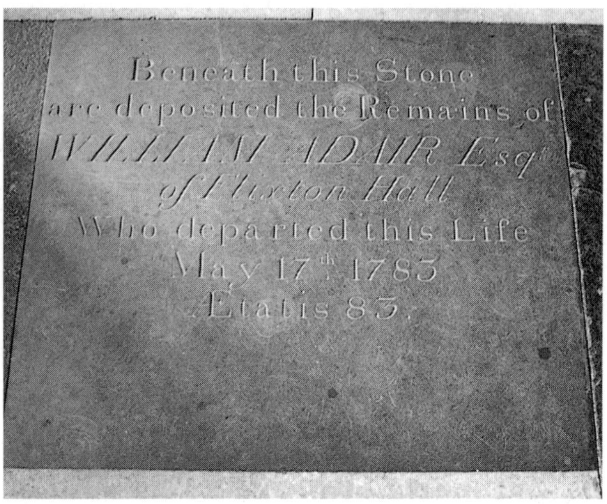

The grave of William Adair in Flixton Church, 2000.

As a memorandum to his will, William Adair made a provision for the executors to provide funds for the purchase of an ensign's commission in the "Foot Guards" for Arthur Brice (born c. 1768?), his grandson, either on his sixteenth birthday or as soon as it could be secured. Young Arthur apparently didn't enter the army through a Guards unit, though. The only early 1780s listing found is for an "A. H. Brice," commissioned as an Ensign in the 85th Regiment on 26 February 1783. After that regiment was disbanded in 1784, it seems Arthur was commissioned as Ensign and Capt. Lieutenant in the Coldstream Guards on 6 April 1785 (Guards officers had a double rank system). He was promoted to Lieutenant and Captain on 27 October 1790 (with army rank of Major dating to 1 March 1794), and served as Adjutant from 22 January 1794 until 1 December 1795. Arthur's final promotion was to was Captain and Lt. Colonel on 2 December 1795. Wounded and taken prisoner on 14 March 1801 in Egypt, he died of his wounds two days later.[26] Arthur had married Ellinor Dupre, and produced a single child, aptly named Alexander Adair Brice.

Alexander Adair carried on in a manner that would have pleased his uncle, keeping both the agency business and the Flixton Estate thriving. During the Napoleonic Wars, Alexander was in a position, both professionally and financially, to help oppose any invasion of England the French might attempt. As its Commanding Officer, he raised the 9th (South Elmham) Troop of Suffolk Yeomanry Cavalry in 1805. For some time, the unit was encamped in the grounds of Flixton Hall's park, and was a point of great pride for Alexander. In 1813, Martin Archer Shee painted a heroic portrait of Alexander as its Colonel, with a number of his troopers in the background.

However, according to the terms of William Adair's will, Flixton Hall and its estate were only to be Alexander Adair's for the term of his life. Therefore, when Alexander died (issueless) in 1834 at the age of 91, the estate passed to his cousin William Adair, of Ballymena, who was then 80 years old. This William Adair was commonly called "Hugh" to differentiate him from the others by that name in the family.

"Hugh," the grandson of Capt. William Robert Adair, had held a Captain's commission in the 25th Regiment of Foot (later the King's Own Scottish Borderers), and had been present at the Siege of Gibraltar in 1781.[27] William Adair was the agent to the 25th Regiment's Colonel, Lord George Lennox, during the time of "Hugh's" service. Therefore, it is probable that William played an important part in "Hugh's" career in the 25th Regiment.

In 1783, "Hugh" inherited a "valuable" gold snuff box engraved with a ciphered "W-A," and a "gold watch made by Graham (of London) with the chain and seals thereto affixed" from William Adair, since they had the same name, and were obviously very close.

In 1784 "Hugh" married Camilla Shafto, the daughter and heiress of Robert Shafto[28] of Benwell, (Northumberland), Esq., bringing the Shafto[29] name and wealth into the Adair lineage. Since "Hugh" married an heiress, the arms of her family were borne "in pretence" in the center of his arms' shield (see page 28, bottom, left).

"Hugh" died in 1844 at the age of 90, and the estate passed to his son Sir Robert Shafto Adair, once a Captain in the 1st Life Guards. Sir Robert was created a Baronet in 1838 for successfully conveying Sir Francis Burdett to the Tower of London (for incarceration) through riotous conditions. Shortly after Sir Robert inherited Flixton Hall from his father, he began to build upon and improve the house, which had apparently fallen into a state of disrepair.[30] To execute these alterations, he employed Anthony Salvin (1799-1881), a renowned architect primarily known for his "Gothic Revival" designs.

Salvin was responsible for the addition of a new wing of kitchen offices on the East Side of the house, along with some improvements to the main building. These included a redivision of some interior spaces, the securing of the foundations and a replacement of the old windows and frames. Since

Alexander Adair, in the uniform of Captain of the 9th or South Elmham Troop of Suffolk Yeomanry Cavalry, engraved by C. Turner after Martin Archer Shee, 1813. The original painting hung in the great hall at Flixton, and was destroyed in the fire of 1846 (ILN, 19 December 1846) The British Museum, copyright reserved.

Flixton Hall, from the northeast, 1823. Engraved by W. Tombleson for J.P. Neale's Views of the Seats of Noblemen and Gentlemen in England, Wales, Scotland, and Ireland, *Volume VI.*

A Presentation of a Rose and Sword to the Lady of the Manor of Flixton, (1838), a watercolor by John Sell Cotman (1782-1842). In a letter to his son, the artist describes this whimsical scene, set on the roof of Flixton Hall, as "One entirely my own in every sense of the word, being very gay and gawdy [sic] and full of frippery." Cotman had been atop Flixton Hall for the preliminary sketch work in 1832. It may be that the weapons carried by these imaginary figures were based on those on display in Flixton's great hall, three of which appeared in the 26 December 1846 issue of the Illustrated London News. *Most interesting is the musket carried over the right shoulder by the figure in the center, which may have been one of the Royal Welsh's Long Land Patterns. Another rooftop view of Flixton Hall , sans figures, is in the collection of The National Gallery of Canada in Ottawa. Courtesy of the Norwich Castle Museum & Art Gallery.*

Salvin's new wing offset the almost perfect symmetry of the original construction, criticism was directed towards it. An article entitled "A Few Fugitive Recollections of Flixton Hall" that appeared in the 23 December 1846 issue of the *Bury & Suffolk Herald* stated;

> This once beautiful, and age-honoured pile of majestic brickwork, had, from time to time, been 'shorn of its fair proportions' by the circumstance of sundry blockings up, and a succession of other architectural disfigurements, chargeable on the depraved taste of those by whom they were originally inflicted.

Towards the end of 1846, the new wing was mostly, if not fully completed, and the bulk of the work was being done in the main part of the house. Large numbers of workmen were employed in removing the old windows and their pedimented frames and replacing them with others of heavy rectangular Caen stone. The house was then open to the elements, so a large quantity of moveable furniture, books and paintings had been piled into an interior room to protect them.[31] Since Flixton Hall was uninhabitable in this state, the Adair family remained at their home in St. James's Square, London (likely Adair House, where they spent most of their time anyway), leaving six domestics behind with the workmen.

Disaster struck early on the morning of 12 December 1846. About one o'clock, a boy passing on the Bungay road noticed flames issuing from the building and raised the alarm. Immediately, some of the furnishings, papers and other valuables were pulled from the building, and messengers were sent to Bungay and Harleston. Three fire engines arrived about three o'clock, by which time the west wing and the central part of the house were fully engulfed. Witnesses described the streams of glowing molten lead pouring from the roof. To try to save the east wing of the mansion and the new offices, Mr. Balls, manager of the Bungay fire engine, positioned himself on the roof at this junction, and for seven hours directed the water hose on the flames from that vantage.

Two early Adair bookplates, the first of William "Hugh" Adair (not the army agent), from a c. 1750 volume formerly in the Shafto library. This 18th century library was moved to Flixton Hall from the seat of the Carter-Wood family in 1939. Note that the arms of the Shafto family are borne in pretence on the arms of the Adair family. The second bookplate bears the arms of Sir Robert Shafto Adair, 1st. Baronet, and is from an early 19th century volume formerly in the library at Flixton Hall.

That morning, an electric telegraph message was sent from Norwich to Sir Robert in London which read;

> Sir,
> It is my painful duty to inform you that early this morning, one O'Clock, Flickston was discovered by one of the night men to be on fire about the center of the Hall. 3 engines were kept constantly playing on it, & we have saved all the new part & part of the east wing, also that part occupied by Mr. Fisher, & its contents, but every thing in the Saloon was destroyed. I think it advisable for you to hasten down. I have not been able to ascertain how it originated, but fear some of the workmens fires were not sufficiently extinguished last night. With sincere commiseration for this most unfortunate accident
> I remain Sir,
> Your most obetdt. Servant,
> Geo. Durrant

Sir Robert Shafto Adair (1786-1869), 1st Bt., photographed by Clare Algar c. 1860-69 in the enclosed colonnade, or conservatory, of Flixton Hall. Courtesy of Glenn Earl.

Sir Robert immediately passed the telegram on to Anthony Salvin, and the next day they proceeded together to Flixton to inspect the damage. What they found must have appalled them—about half of the house was destroyed. Thanks to Mr. Balls of Bungay, the east wing and the new offices were indeed saved, but the center part of the house and the west wing were completely gutted. Even the massive support beams, which were about 18 inches square, had fallen to the ground, charred and in pieces. Of the gutted part of the house, some of the walls were so severely burned and cracked that they had to be pulled down before it was deemed safe to begin removing the rubble on the floor. One wall on the south side was so damaged that it fell outward on its own shortly after the fire was extinguished.

While it is clear that not every valuable and possession in the house was lost, a significant portion of the furnishings and paintings were. This included the contents of the salon on the second floor, known to have housed many treasured paintings and antiques. Amongst the losses was the portrait of General Huske, Colonel of the Royal Welsh Fuziliers, in addition to portraits of other naval and military notables of the eighteenth century.[32] The house and its contents were insured by the Norwich Union Fire Office, and the losses incurred were thought to exceed 60,000 Pounds, by some accounts. Almost immediately, Sir Robert announced that he would completely restore Flixton Hall, under the direction of Salvin.

During the spring or summer of 1847, Salvin did a watercolor sketch of the damaged mansion from its side, looking directly into the west wing. Floor plans for the restoration survive in the archives of the Royal Institute of British Architects, and show the walls that needed replacement in red. These documents serve to illustrate the damage done to the house, and help determine which original elements remained intact.

Among the very few surviving views of Salvin's Flixton Hall are six glass plate images found in the 1980s above a shop in

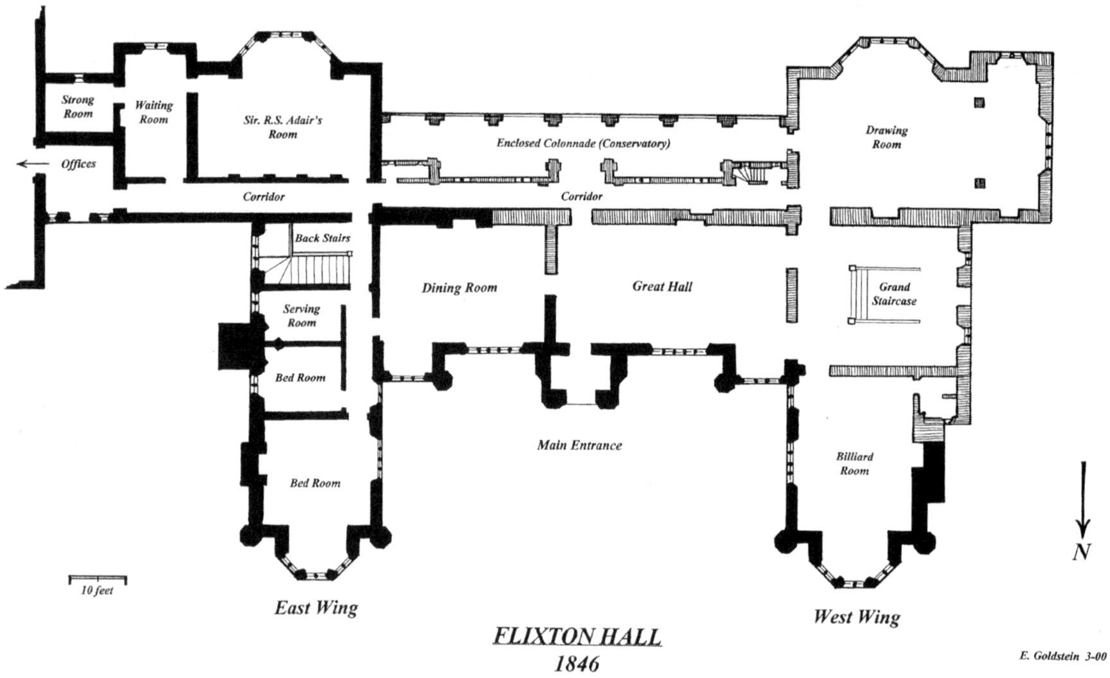

Floor plan of Flixton Hall c.1846, after Salvin. Those walls destroyed by the fire are hatched. To the upper left are the new offices added by Salvin just before the conflagration.

The ruins of Flixton Hall shortly after the fire, by Salvin, 1847. This view is from the west and shows the arches separating the grand staircase area from the great hall, and the void above, which was formerly the salon, home to many of the paintings lost. Salvin had obviously not yet replaced the pedimented window frames in this part of the house. Atop two of the pinnacles are seen decorative iron finials bearing a "flag" pierced with a gothic "A." These were in place until the house was demolished in 1952-53. Courtesy of Lady Darell.

DESTRUCTION OF FLIXTON HALL BY FIRE.

We regret to record the entire destruction of the fine old "ancestral home," situate at a short distance from Bungay, in Suffolk. The details of the catastrophe are as follow:—

At about two o'clock [illegible] Sunday morning, the ringing of the church bells and t[illegible]d cry of "Fire!" aroused the inhabitants of Bungay, and, upon the locality being ascertained, one of the town engines and the powerful one of the silk mills at Ditchingham, were, after a short delay in procuring horses, despatched, followed by hundreds of persons, who were quickly joined at the scene of the fire by an engine from Harleston; but so rapid had been the progress of the flames, and so entirely had they obtained the ascendancy, that comparatively nothing could be saved; and the house, with its furniture (some of which was very costly), its pictures. fine old china, &c., fell a sacrifice to the flames. The fire was not subdued till late on Sunday evening.

Another account states:—Late on Saturday night a dreadful fire broke out atFlixton-Hall, the residence of Sir Robert Shafto Adair. It appears that about twelve o'clock at night, a boy was passing along the road adjoining the parks when his attention was called to a great body of flame issuing from the window, of the Hall. He immediately gave an alarm, and aroused the servants. Shortly afterwards, the engines arrived and began to play on the burning pile, but not to much effect, as the fire had got such a hold as to defy all their efforts to stop it. The mansion was destroyed, with all its valuable and ancient pictures (one worth 1,000 guineas), and costly furniture. Nothing was saved from the flames but a few bolsters and pillows. The family were absent, and there were only six domestics in the house at the time.

A messenger was despatched to Norwich, in order to communicate, per telegraph, the intelligence to Sir Shafto Adair, in London; and, in the evening, Mr. A. S. Adair, the elder son, arrived.

The bare walls are standing, the whole being completely gutted. The Hall has been under extensive repair the last half-year, and a great number of hands have been employed: many of their tools were consumed. The loss cannot at present be estimated, but it is thought that £40,000 will not cover the damage.

Flixton is stated to have been built about the year 1615, and continued to be the residence of the ancient family of Tasborough, from the site of its erection to the middle of the present century, when it was purchased by the late William Adair, father of the present owner, created a Baronet in 1838. The building has been attributed to Inigo Jones; but it has the earlier characteristics of shafts, like the ornamented chimneys of the Elizabethan style, at the bayed projections of the wings and centre. The whole edifice had a vast number of windows, all of them pedimented. The doorway is arched, and flanked by coupled columns, supporting a pediment. The pillars are placed on pedestals (ornamented with lozenges), elevated on the base. The building is battlemented; and the corners of the wings have buttresses, rising similar to chimney-shafts. The whole appearance is, or rather was, noble; and the building was a good specimen of the mixed style prevalent in the seventeenth century, but not a highly enriched one,—not so highly, at least, as many of the old halls to be met with in Suffolk and Norfolk. We have engraved he principal or northern front.

Of the artistical treasures of the mansion we find the following account in Raw's Pocket-book, of some years since:—

"In the hall are busts of the Right Hon. C. J. Fox, Lord Keppel, and General Wolfe. In the staircase, above the door entering the saloon, is a fine bust of Inigo Jones, who built the house. The saloon contains thirty-four paintings, by various masters, among which are the following:—A Madonna and Child; Saint Peter and the Angel; Fruit and Flowers, by Van Os; Saint Mark's, at Venice, by Canaletti; Landscapes, by Tillemans; Sea Pieces, by Vandervelde, &c. The library contains a choice collection of books; a portrait of W. Adair, Esq., with his groom and two horses; a portrait of the Duke of Bolton's famous horse Sweepstakes: both this and the saloon are excellent rooms. In an adjoining bed-room is a portrait of General Huss, of Elling House; a Turkish Lady; a Battle Piece; Joseph and Potiphar's Wife, &c., all very finely painted. In the other bed-rooms are the following paintings:—St. John the Baptist's head; St. Agnes; Mark Anthony and Cleopatra; a Sleeping Venus; Lucretia, &c. In the dining-room is a portrait of the present proprietor of this mansion, whose pleasant countenance confirms the public report of his politeness and urbanity. In the drawing-room are judiciously collected together the portraits of the late Duke of Richmond, his father, his mother, and sister, by Sir Joshua Reynolds; the late Duke of Cumberland; Sir Charles Saunders, by Sir Joshua Reynolds; Lord Anson, Keppel, Lord Albemarle, General Hughson, Lord George Lennox, General Napier, &c."

News of the fire at Flixton Hall as it appeared in the 19 December 1846 issue of the Illustrated London News. *The engraving of the Hall was made from sketches taken a few years before the addition of the new wing.*

CARVED DOORWAY AND BLACK JACK, AT FLIXTON-HALL.

FLIXTON HALL, SUFFOLK.

AST week we gave an exterior view of this fine old mansion, and an account of its recent destruction by fire; we are now enabled to add some few more particulars of its history, together with Engravings of some of the characteristic features of the interior and its furniture, from original sketches, taken about three years since, and which have not hitherto been engraved. The house, although an interesting example of its peculiar style, and an important building in itself, was little visited by the topographer or antiquary, and is but slightly noticed in county histories. This may be owing to its situation in a part of the country where little traffic occurs. It stood in a fertile and beautiful park, watered by the river Waveney, the grounds in front, and the plantations surrounding it, were of the most luxuriant character.

Abont the time of Henry III. an Augustine Nunnery was established here, by Margery, widow of Bartholomew de Creke, and is another instance of the excellent sites chosen in old times for the building of religious houses, with wood, water, and pasture land, in convenient proximity. The name of Flixton is supposed by some to be derived from St. Felix, the first Bishop of the East Angles. The Nunnery was dissolved in 1528, and intended to be given to Cardinal Wolsey, but, on his declining the offer, it was granted to John Tasburgh, in the failure of whose descendants it passed into the Wyburn family, and was purchased of them by Mr. Adair, about the middle of the last century.

The period at which the late Hall was erected is not satisfactorily settled—it is called Elizabethan by some; and it certainly possesses features of the peculiar style of architecture so called; but others confidently say that it was erected in 1615, in the style called "Inigo Jones's Gothic." The truth seems to be, that Jones was employed to re-model the old house; at least, this is the traditional story; and that to his design the front of the building is owing. Originally the house was surrounded by a moat, but this had been filled up long since. The ground plan of the front, with its projecting wings, and slightly projecting doorway, forms the letter E, which was so fashionable a style of construction in the reign of Elizabeth, as a compliment to her, that it may be a reason for imagining the substructure older than Jones's time. The crocketed pinnacles were characteristic ornament to its roof, which was perfectly flat and battlemented, and from whence an agreeable view of the surrounding fertile country was obtained. The Hall was a spacious room, filled with antique furniture and armour. The large carved fire-place was furnished with logs in the style of the past ages, and many interesting family pictures hung around the walls. A flight of stairs led from the Hall to the Saloon, the entrance door of which forms the subject of our Cut, and was elaborately carved in oak, with scroll-work foliage and figures, in the style of the *Renaissance*, but rather more chaste in its design than usual. Above it was placed a bust of Inigo Jones, whose connection with the house, as a builder or re-modeller, we have noticed. The suite of rooms leading from the Saloon, as well as that apartment, had been so altered for modern convenience, that nothing of a striking character had remained from its original state. The furniture and pictures were valuable, one painting being valued at a thousand guineas; the cabinets were also costly in construction, and of ancient workmanship. But the most remarkable piece of furniture was the folding-table here engraved, which was richly carved, and might be placed flat against a wall, or used by simply adjusting the three legs, when it was wanted. Two of the legs were immoveable, on a carved cross-bar, which turned on a pin; the third leg moved on a hinge in the centre of the table, and, being brought into its proper place, the table top was lowered upon it, and the upper part of this leg fitted into the lion's mouth, and secured by the pin which hung beside it. Like too many of our ancient mansions, Flixton did not possess many relics of ancient times, which modern manners dispense with; but one curious thing remained, and was treasured with care; this was the "Black Jack," or leathern jug, in which ale was dispensed in the Hall on great festivals, when the servants and tenantry were assembled. That at Flixton was one of the largest known to exist, and was capable of holding eight gallons; it is figured in our Cut of the Entrance to the Saloon, together with some few pieces of Defensive Armour, which was preserved in the late building as an ornament to the walls of the Great Hall.

A follow-up story from the 26 December 1846 issue of the Illustrated London News. *The engraving shows the carved doorway into the salon on the second floor with some of the arms and armor from the great hall on the floor below. The huge "blackjack" had been at Flixton Hall "since ancient times," and survived the fire. It is still a cherished family heirloom.*

the town of Diss occupied by a photographer named Clare Algar during the 19th century.[33] Although they are undated, these images probably date from c. 1860 to 1870, and suggest that the criticism leveled at Sir Robert for the new offices was not entirely baseless. Salvin altered the exterior of the original house little, only remodeling the front entrance and adding a square cupola to the roof at the center of the building, in addition to the aforementioned new window frames. The new offices jutted out to the east, spoiling the symmetry of the original design. Additionally, these offices were nothing like the main house in architectural style, being far plainer and much lower in height, with gabled roofs.

The interior was very tastefully finished with massive joinery reflective of its original construction, with some of the walls on the ground floor covered in geometrically carved hardwood panels. Many considered it one of Salvin's most lavish interiors.[34] For the years of work Salvin put into restoring Flixton Hall, Sir Robert was billed 29,100 Pounds.[35]

Sir Robert seems to have further remodeled the house periodically and, in 1863, he had the keystones of the arches at the West end of the great hall carved with his initials and the date of the alteration. At some point between the 1860s and 1873, a heavy openwork moulded brick wall was put up surrounding the mansion and its gardens, with a large iron gate centered on the front entrance of the main part of the house. These alterations seem to have been executed by Salvin too.

In 1869, Sir Robert died and was succeeded by his son, Sir Robert Alexander Shafto Adair. Raised to the Peerage in 1871 as Baron (Lord) Waveney of South Elmham, Sir R. A. S. Adair had led a colorful life. In addition to managing the Adair estates in both England and Ireland, he was a Lt. Col. in the East Suffolk Militia Artillery from 1853 to 1881, honorary Colonel of the same from 1881 until 1886, the Lord-Lieutenant of County Antrim, and a Member of Parliament for Cambridge for a while. However, his most high-profile role was as aide-de-camp to Queen Victoria from 1857 until his death.

Lord Waveney outlived his wife by almost sixteen years, and died childless in 1886. Therefore the peerage became extinct, and the baronetcy passed to his younger brother and heir, Sir Hugh Edward Adair, 3rd Bt., Ipswich's Member of Parliament for a time, and a barrister.

Flixton Hall from the northeast, photographed by Clare Algar, c. 1860-70. Taken from almost the same vantage as the 1823 engraving for Neale's Views, *it, and the following two photographs, are probably the earliest of the mansion. The cupola at the center of the building was added by Salvin. The man in the center with the top hat and the gun under his arm is probably the estate's head gamekeeper. Courtesy of Glenn Earl.*

Flixton Hall from the northwest, taken by Clare Algar c. 1860-70. Salvin's offices of the 1840s are at left, and the heavy brick wall surrounding the house and gardens had not yet been added. Courtesy of Glenn Earl.

Flixton Hall from the southwest, c. 1860-70, taken by Clare Algar. Salvin's cupola can be clearly seen, as well as the colonnade that was closed-up between 1753 and 1787. The woman in the white dress may be Lady Adair. Courtesy of Glenn Earl.

The great hall and grand staircase, c. 1863-71. This staircase is the one installed by Salvin in the late 1840s, and is far more restrained than that installed 20 or so years later. Arms displays, including four of the halberds and what is probably one of the Royal Welsh infantry hangers, are seen at left. The elegant woman on the staircase is probably Lady Adair (d. 1871), wife of the future Lord Waveney (elevated in 1873). A beautiful white marble statue of her stands in its own memorial chapel in Flixton Church. Courtesy of Lady Darell.

Flixton Hall, Suffolk. The Residence of Sir Robert Alexander Shafto Adair, Bart. Since Sir R.A.S. Adair inherited the baronetcy in 1869, and became Lord Waveney in 1873, this engraving dates from between these years. This is the earliest view of the heavy brick wall seen there today. Courtesy of Glenn Earl.

Lord Waveney (1811-1886) by "Spy" (Sir Leslie Ward), from Vanity Fair *magazine, 2 April 1881.*

From 1888 to 1892, Flixton Hall was again remodeled, this time by an architect named F. Wade, under hire of Sir Hugh. A number of 20th century writers have credited the final appearance of Flixton Hall to Salvin, when in truth, it was largely owing to Wade. The most drastic of these new alterations was the replacement of Salvin's office's on the east side of the mansion. Perhaps Sir Hugh thought criticisms of the 1840s were true, and sought to correct the problem. To the front, Wade's new offices reflected the main part of the house in style and profile, in addition to being significantly more extensive than the old offices.

Many exterior embellishments were added, including an ornate new entrance to the main hall. The rear of the house received a stepped string course of sculpted stone skirting many of the rooflines, and the enclosed colonnade was converted into a conservatory with two skylights. In place of Salvin's cupola in the center part of the main house was built an elaborate stone clock tower, resembling a tiered wedding cake, modeled on the tower of Christ Church, Oxford. On the west wing, a highly ornate entrance portico was added, with staircases leading to both the gardens and the front of the house.

Significant changes were also made to the interior, the most important of which was the redivision of the central areas between the two wings of the main house's ground floor. Originally, this large area was divided between the great hall, which was about 42 feet by 20 feet, and the dining room, which was about 25 by 20 feet. The combination of these made the new great hall an impressive 73 or so feet long and about 20 feet wide. A new triple archway replaced the wall separating the old dining room from the east wing of the main house. This new archway was almost identical to the newly remodeled one (formerly 1863-dated) on the other side of the great hall. Replacing the more subdued staircase that Salvin had put in after the fire, one made of highly intricate woodwork, gracefully incorporating the stylized initials of Sir Hugh was installed.

Sir H. E. Adair died in 1902 and was succeeded by his son Sir Frederick Edward Shafto Adair, 4th Bt. Sir Frederick was a gentleman of leisure, and had served as a Captain in the Rifle Brigade. He additionally served as a Justice of the Peace for both Norfolk and Suffolk, and also as High Sheriff of the latter county. Sir Frederick spent much time abroad, resulting in a number of publications, concerning places at which he had spent time, such as *Sport in Ladakh* and *A Summer in High Asia*. His contributions to Flixton Hall were the additions of a rock garden of Derbyshire stone and a series of gardens based on themes concerning the four seasons. Sir Frederick was also an accomplished naturalist and photographer, responsible for many of the fine photographs of Flixton Hall and its gardens.

When Sir Frederick died unmarried and childless in 1915, both Flixton Hall and the baronetcy passed onto his brother, Sir Robert Shafto Adair, 5th Bt. Commonly called

Flixton Hall from the north, early 20th century. This picture shows the mansion in its final state after the renovations of 1888-92, showing the new offices at left, the new entrance and tower in the center of the main house, and the grand entryway into the west wing at the extreme right. Courtesy of Lady Darell.

Flixton Hall from the south, c. 1900-10. This photograph provides a good view of the embellishments added to the south facade by Wade between 1888 and 1892, and should be compared to the Algar photo of the 1860-70 period (page 34, bottom). What was formerly the enclosed colonnade became the conservatory, where oranges and various plants were grown. A small group of soldiers are seen at center, and may be Artillery Volunteers.

The great hall, looking from the east towards the grand staircase, c. 1907, possibly taken by Sir Frederick Adair. The central part of this view is similar to the c. 1870 view (page 35, top). The dark areas under the arches are where some of the weapons were displayed. Courtesy of Lady Darell.

"Sir Shafto," he was one of the few Adairs of Flixton Hall who had never been a soldier. Sir Shafto was a barrister and, most notably, a director of the Royal Academy of Music.

Sir Shafto's only surviving son, Allan Henry Shafto Adair, born on 3 November 1897, was destined for greatness. Following in the family tradition, young Allan attended Harrow, where Adairs had been educated since about 1780. Upon graduation in 1916, Allan was commissioned into the Grenadier Guards, and shortly afterwards found himself fighting on the Western Front during the First World War. On the eve of his twenty-first birthday, while sipping the rum of Lord Harewood, he signed papers pertaining to his estate, should he be killed. The next day he was hit in the knee by machine gun fire while attacking the village of Preux.

By the eve of World War Two, while instructing at Sandhurst, Major Adair was ordered to immediately take command of the 3rd Battalion Grenadier Guards in Belgium. During W. W. II, he earned the rank of General, the affectionate regimental nickname "General Allan" and was given command of the Guards Armoured Division. General Allan led it very successfully through the European theatre, fighting at the Normandy breakout, the liberation of Brussels, the advance on Arnheim and the crossing of the Rhine. On 9 June 1945, the Guards Armoured Division held its "Farewell to Armour" parade in Germany. Field Marshal Montgomery attended the event and gave a tribute to General Adair of the highest order.

World War II was also a very busy time at the Adair estate in Suffolk. In order to launch raids on enemy targets in Europe, the American 8th Army Air Force was stationed throughout Southern England. At the peak of the war, there were over 100 bases with more than 400,000 men and

Major General Sir Allan Shafto Adair, Bt., GCVO, CB, DSO, MC, JP, DL, during, or shortly after, WWII. Courtesy of Lady Darell.

over 11,500 planes. One unit, the 446th Bomb Group, nicknamed "The Bungay Buckaroos," was stationed at Flixton (Station 125), and in fact, its Servicemen outnumbered the residents of the area. In early November of 1943, Flixton Hall survived a near miss unscathed when an unmanned B-17 bomber, flying on auto-pilot, crashed just south of the mansion, with a large explosion ensuing.[36]

The pleasure of the success and honor earned by General Adair in World War Two was tempered by personal tragedy. Desmond, the General's only son and heir, was serving as a Captain with the 2nd battalion Grenadier Guards. On 10 November 1943, at the Battle of Monte Camino, Italy, Desmond was killed. It was not until June 1945 that his body was recovered, from an unmarked communal grave containing mostly German dead.

Following the war, General Adair served in Greece, and retired from the army in 1947 after his Division was disbanded.[37] He kept very busy, taking over the management of Sir Shafto's farms and estates at Flixton for a short period until they were sold in July 1948.[38] The Flixton Estate included some 2970 acres of land and an annual rent roll of about 3500 Pounds Sterling. It included 21 farms, two public houses, two schools, three village post offices and various houses and cottages, and was purchased by Metropolitan Railway Country Estates, Ltd. It was then further divided, and some of the tenants were allowed to purchase their farms.

The Adairs only retained Flixton Hall and its park, in addition to its Home Farm and Home Woods. Sir Shafto died on 9 October 1949, and was succeeded by General Allan as the 6th Baronet.

The post World War II British economy was very much focused on rebuilding that which German bombs had destroyed, and recovering from the tremendous expenses of the war. Consequently, to generate revenue, tax structures changed, and the brunt of the newly redirected load fell on the upper classes. This was unequivocally responsible for the downfall and destruction of countless magnificent country houses and estates, causing the dispersal of their priceless contents.

Such was the case with Flixton Hall. Even though it had been in the Adair family for nearly 200 years, General Adair was forced into selling it to satisfy the heavy death duties levied on his father's estate. From his 1986 memoirs, *A Guards' General*, we see how General Adair felt about Flixton Hall;[39]

> Flixton Hall was a vast, uncomfortable mausoleum still with no proper central heating.... It was so bitterly cold there in winter that our children had to wear their overcoats when they went from room to room!

It was apparent that by the mid-twentieth century, Flixton Hall was too large, too uncomfortable and too expensive to keep and maintain, especially without a male heir to take over the estate. He was, however,

Men of the Ordnance Section of the 705th Squadron, 446th Bomb Group, on the front wall of Flixton Hall, 1944. From left to right are; top row—Phil "Softy" Schenker, Fred Mahnken & Leonard "Gremlin" Mayer, bottom row—Vincent DeAngeles & Alex Nole. Courtesy of Fred Mahnken, 446th Bomb Group.

Aerial view of Flixton Hall from the northeast, taken sometime during 1944 by an American "Photo Joe" stationed at the U.S. Army Air Corps base at Flixton. Courtesy of Fred Mahnken, 446th Bomb Group.

Aerial view of Flixton Hall from the southeast, also taken in 1944 by a "Photo Joe." This view shows the majestic countryside surrounding the mansion. Courtesy of Fred Mahnken, 446th Bomb Group.

upset about his family home's fate. Reflecting on it later in life, he also wrote in his memoirs:

> It is now used as a cow-shed, and the cows look rather sad about it all. I feel the same way as Flixton had been in the family for over 200 years.

By the end of 1950, now retired, General Adair had set up residence at Raveningham, and was making arrangements for the contents of Flixton Hall to be either removed or sold. Most of the items not kept by the family were put up for public auction between 27 November and 1 December 1950 by the firm of Stanford, Broom & Stanford of Halesworth, Suffolk. The catalog listed 1880 lots, and the library of 7000 volumes is said to have taken a whole day to auction off. The sale netted slightly over 15,955 Pounds Sterling, roughly around $65,000 at the time.

Efforts were made by both the East and West Suffolk County Councils to buy the mansion and the 250 acres of land surrounding it for use as a joint farm institute, but to no avail. The Ministry of Agriculture found the property not well located, and the land unsuitable.

A speculator named Mr. R.G. Lawrence, who claims to have "tried every angle in effort to preserve it",[40] purchased the mansion and its park from General Adair. It seems more likely that Mr. Lawrence was far more interested in profit, since the sum of Flixton Hall's parts were far greater than the whole. The gravel pits to the north of the house were sold, and it is rumored that the price of the roof leads alone, sold as scrap metal, brought the entire cost of the mansion and land back to Mr. Lawrence.[41]

In early June of 1952, after the house had been vacant for almost two years, local newspapers announced "Demolition of

Flixton Hall Due to Start on Monday." Mr. Lawrence justified this by stating that "the place, which has been empty for about two years, is beginning to show signs of dampness. If it stood for another year or two, it would begin to tumble down." Perhaps if the lead from the roof had not been removed and sold off, then water may not have found its way in.

It was further publicized that the outbuildings, and those parts of the main house that were suitable, would be converted for farm use. Mr. Lawrence was well aware that much of the material used in the construction of Flixton Hall was valuable, including the fine interior woodwork, windows, stonework and brickwork. Only a few men were employed in the demolition, which

Aerial view of the remains of Flixton Hall from the northwest taken on 19 July 1979. Just how much of the mansion was removed is evident when compared to the 1944 aerial views. National Monuments Record, copyright reserved.

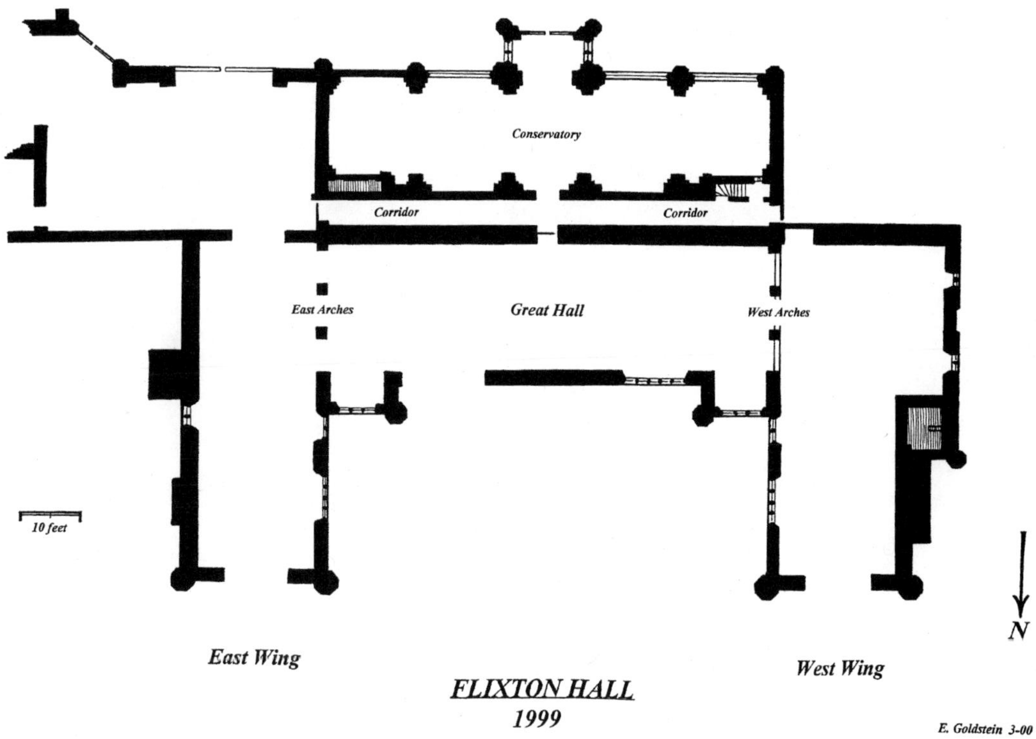

Floor plan of the remains of Flixton Hall, 1999.

took approximately a year to complete. As this valuable material became available, auctions were periodically held on site to sell it off.

Of the fine wood paneling, much of it went to Norwich, where it was used to furnish the "Flixton Room" at the old "Samson and Hercules Ballroom."[42] American soldiers stationed in East Anglia during the war frequented this famous establishment. Currently, the whereabouts of the grand staircase is unknown.

The main part of the house, with its many walls dating to the beginning of the seventeenth century, was carefully dismantled from the top down. When the demolition and conversions were complete, all that was left of the original mansion were most of the outer ground floor walls, stripped of their embellishments. The polygonal bays at the ends of the east and west wings were removed and converted into garage entrances for the farm machinery now housed there. The entrance portico leading into the great hall was removed and its opening bricked up. The ornate stone staircase and entranceway on the side of the west wing were completely removed, along with the entire drawing room. Most of the interior walls were removed, and all of the fireplaces were bricked up. Since everything above the first floor was gone, a sloping three-sectioned corrugated metal roof was installed.

To the north front, the new offices built c. 1888-1892 were completely removed. Most of the offices towards the rear of the mansion, like the carriage houses and stables, have survived, but were stripped of most of their trimmings. Miraculously, the dovecote,[43] known to be Jacobean, survives untouched from the earliest days of the Tasburgh's ownership of the mansion.

Although a terrific amount of irreparable damage has been done to Flixton Hall, enough of it has survived to tell vividly of its history and former grandeur. To see it as it appeared in 1999 is to lament its loss deeply. Surely, an act of such vandalism inflicted on so fine an old house would not be allowed today....

The approach to the front gate of Flixton Hall, March 1999.

The remains of Flixton Hall from the northwest, taken February 2000. This was taken from about the same vantagepoint as Sandby's 1752 watercolor (see page 22).

The remains of Flixton Hall from the southwest, taken in March 1999. This view roughly approximates the vantage point in the early 20th c. view where the ceremony is taking place (see page 37, bottom).

The Lineage of Flixton Hall in the Adair Family 1753-1950

1- William Adair, 1753-1783, to his nephew.
2- Alexander Adair, 1783-1834, to his cousin.
3- William "Hugh" Adair, 1834-1844, to his son.
4- Sir Robert Shafto Adair, 1st Bt. 1844-1869, to his son.
5- Baron Waveney, Sir Robert Alexander Shafto Adair, 2nd Bt., 1869-1886, to his brother.
6- Sir Hugh Edward Adair, 3rd Bt. 1886-1902, to his son.
7- Sir Frederick Edward Shafto Adair, 4th Bt. 1902-1915, to his brother.
8- Sir Robert Shafto Adair, 5th Bt. 1915-1949, to his son.
9- Maj. Gen. Sir Allan Shafto Adair, 6th Bt. 1949-1950 (died 1988).

The early 17th century dovecote at Flixton Hall, March 1999.

The Weapons Collection

At the time General Adair was making arrangements for the dispersal of Flixton Hall's contents, he came in contact with William Keith Neal, and arranged for a private sale of the weapons used to decorate the walls flanking the great hall on the first floor.

William Keith Neal (1905-1990) has come to be regarded as one of history's greatest antique firearm collectors and authors. Neal's interest in weaponry began early in childhood, and his training in photo portraiture served to make him an expert gun photographer. During World War Two, Neal worked for the Ministry of Economic Warfare, which sent him to the United States to sell antique firearms, many of which had been donated from Britain's great country houses. In turn, the proceeds of these sales went towards the purchase of modern weapons for Britain's war effort. This was quite dangerous work, as Neal once found himself adrift in the Atlantic for fourteen days after his ship was torpedoed from under him.

William Keith Neal (1905-1990), photographed in 1950, the year he purchased the arms displays from General Adair at Flixton Hall. Courtesy of Mrs. Diana Berry.

On 12 October 1950, Neal made the more than 200 mile trip from Bishopstrow, his home in Wiltshire, to Suffolk to see the collection. He apparently bought the whole lot then and there, with the exception of a few, probably minor, pieces that were sold in the following month's house sale.[44] Additionally, Neal purchased most, if not all, of the fine sporting arms in the Adair gun cabinet.[45]

At this time, Colonial Williamsburg was very actively looking to furnish its many historic buildings with proper period antiques. One of its prime attractions, the Powder Magazine, required a large quantity of antique firearms and other weapons to recapture its appearance during the early 1770s. To accomplish this, relationships were formed with many of the prominent dealers and experts of the period. One of the most fruitful was that formed with William Keith Neal, whose position with the Ministry for Economic Warfare may have introduced him to Colonial Williamsburg during the war years.

Neal immediately recognized that the collection of weapons he had purchased would be perfect to help furnish the Powder Magazine. Three days after his initial visit to Flixton Hall, on 15 October 1950, he excitedly wrote to M. W. Thomas, Jr. at Colonial Williamsburg;

Now at last I can report some good news regarding some of the items you still need for the Armoury (Powder Magazine). I have just succeeded in buying a most remarkable family collection of arms that includes some excellent examples of early 18th. century sergeant's halberds and also infantry swords of the same period. There are also a few of the rare early dated muskets as supplied to the Guards with 46 inch barrels and have the contractors name such as Jordan, G. R., Tower and the name of the regiment engraved on the barrels such as Royal Welsh Fuziliers etc. These muskets match up with the early dated long Dragoon Pistols which I sent you for the Armoury, and they are the only stand of these arms I have ever been able to acquire. They have been in the possession of the family from whom I purchased them ever since the days of their use, and have only been sold to me because the mansion has to be sold to pay death duties. The family have been associated with The Grenadier Guards for an unbroken succession of many generations and this accounts for the preservation of these arms.

There are seven sergeant's halberds in fine clean state, one spontoon,[46] and two groups of infantry swords which appear to be all of the same pattern and a certain number actually have their scabbards. All of these arms are hanging in the long gallery very high up and until I have got them down I cannot send you full details but there will be about twenty infantry swords of the 1740-1760 period which is I think just what you want...

So far I have only collected some of the arms from this collection, the majority have to be taken down with ladders from the high walls and I am making arrangements to fetch the rest in the course of the next ten days. When I have all of the arms here I will sort them out and send you details of those which I think will interest you, also I shall photograph some of them so that you will know exactly what they are like. I am most excited at getting this collection which I regard as one of the most important I have ever had the good fortune to acquire.

Neal was diligent about returning to Flixton Hall to collect the remainder of the arms which he had not removed on 12 October. He followed up with another letter to M. W. Thomas, Jr., on 2 November 1950;

At last I can report good news about the arms you still want for the magazine. I have taken photographs which I enclose of a number of the weapons that I have found.

No. I. photograph shows the hilt of an infantry hanger c. 1742. It has the correct curved blade, cast brass hilt with brass grip embossed with the arms of the regiment, The Royal Welsh Fuziliers....I have been most fortunate in buying a complete trophy of these from one of the oldest families in England connected with the Army. These arms have been in their possession for over 200 years and to have found a complete "run" of such attractive ones as these is really more than I had hoped as I quite expected to have to find them in ones and twos and then with so many variations they would not look so well in your magazine.

I have twenty six all told of these swords all exactly alike. Of this number 21 are perfect, nine actually have their original leather scabbards. Of the remaining five three have had the quillon broken off, and two have lost the tips of their blades. These five are still quite desirable and the damage is not very noticeable and I think genuine battle scars.[47]

Photo 2. Shows a sergeants halberd. It has the initials G.G on it (Grenadier Guards). This halberd has its original shaft with iron shod point. I have six other sergeants halberds identical to this all from the same stand of arms and all in fine state, the others do not have G.G on them but they are exactly the same. The spear head unscrews from the blade as it should do. They have the blade passing through a slot in the ferrule, they are of uniform thickness and not sharpened. They are the best specimens I have ever seen and nothing better exists even in the Tower. The shafts are in good shape excepting two which have some worm holes which have been treated they are not weak and the few worm holes are not serious....

I do not know if you have yet received the Grenadiers musket with 46" barrel but I can now give you the news that in this same family collection of arms I have purchased a stand of these early muskets, all

Six of the brass hilted Royal Welsh Fusilier hangers, four of which have their scabbards, leaning against what is probably a garage door at Neal's home, Bishopstrow House. This picture was attached to Neal's letter of 2 November 1950 offering the items from Flixton Hall to Colonial Williamsburg. The Colonial Williamsburg Foundation.

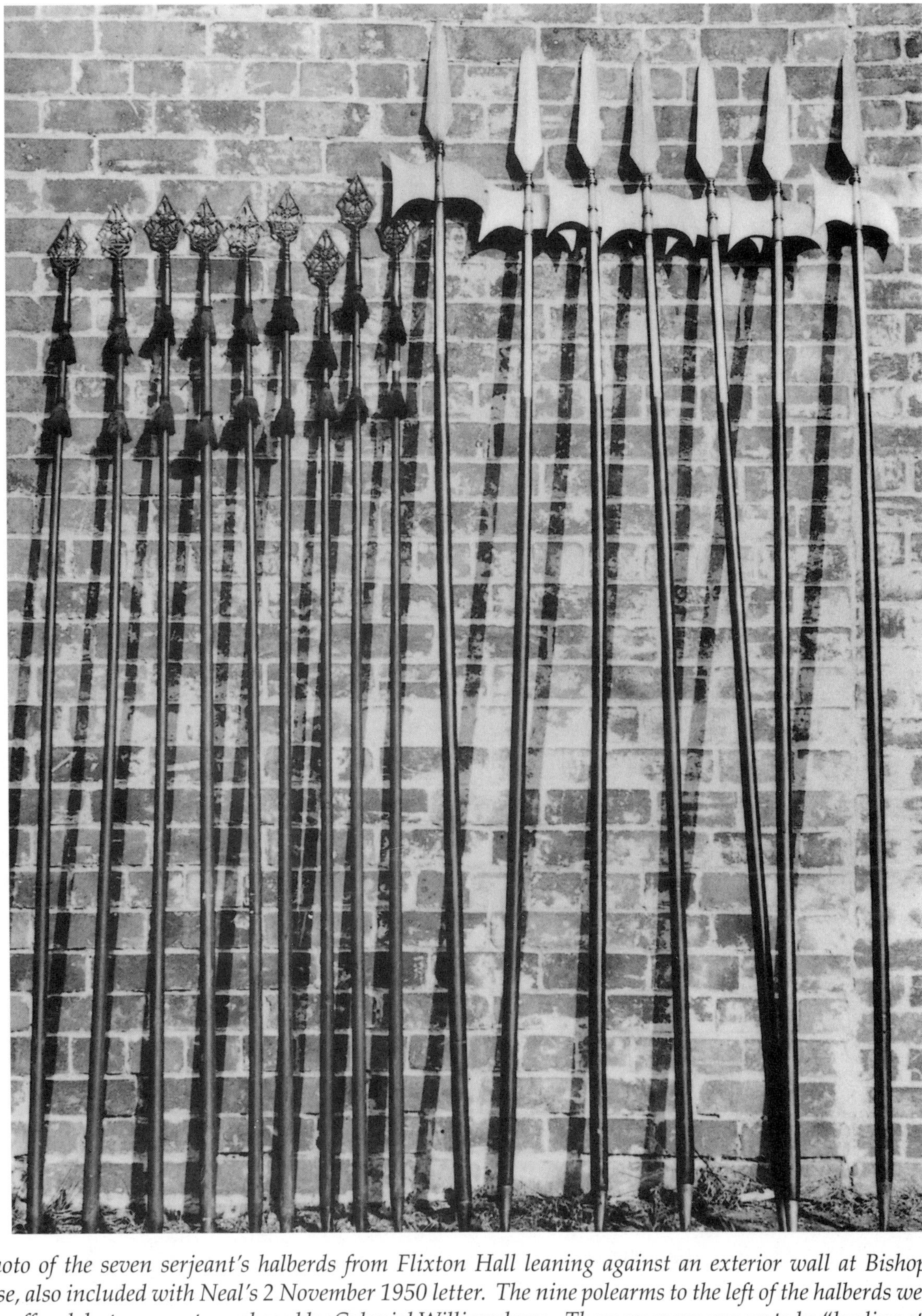

A photo of the seven serjeant's halberds from Flixton Hall leaning against an exterior wall at Bishopstrow House, also included with Neal's 2 November 1950 letter. The nine polearms to the left of the halberds were also being offered, but were not purchased by Colonial Williamsburg. These weapons appear to be "leading staves," as used by officers during the 17th century, or 19th century decorative reproductions. Although they may have been obtained from Flixton Hall, they would not have been of use to any of the regiments to which the Adairs were agents. Perhaps these are the "pikes" Lady Darell recalled in 1999 as being part of the arms displays. The Colonial Williamsburg Foundation.

The 18th century Powder Magazine at Colonial Williamsburg, built in 1715, and restored to the way it appeared during the early 1770s. The Colonial Williamsburg Foundation.

with 46" barrels and dated between 1730 and 1760. Some have wooden ramrods and some have iron, all have their own bayonets. Nearly every one is marked with the regiment Royal Welsh Fuziliers or else the Welsh Arms (Prince of Wales feathers and the words Ich Dien). They are the finest collection of these muskets in England, even the Tower have nothing like so many as I have, it is the most exceptional good fortune that I have been able to buy them. Every musket is perfect and has the heavy moulded brass mounts, swelled fore-end, large brass top escutcheon on hand grip. The dates vary from about 1728 through 1760 and all have the G.R. Crown and either the contractors name like Jordan, Farmer, etc. or else the word Tower. The one Grenadier musket that I have sent you was only the second of these I have ever seen out of captivity so imagine my pleasure in getting a whole stand. Although all these muskets are sound and in perfect order quite a number have old worm holes in the stocks. There is no active worm as the stocks have been treated, and they are not weak, though several of them have had severe attacks.... I can offer you a stand of twenty of these Grenadier muskets all with 46" barrels and their bayonets....

Almost immediately, Colonial Williamsburg purchased the twenty muskets and the seven halberds, all of which arrived in early 1951. They decided to delay purchasing the infantry hangers until early 1953, since Neal had reserved them specifically for Williamsburg. A Pattern 1730 musket and a Pattern 1756 musket were Colonial Williamsburg's final purchases of arms from Flixton Hall, arriving in early 1954. Both muskets were marked to the Royal Welsh Fuziliers.

It is truly fortunate for arms students that Neal decided to place the overwhelming bulk of this most remarkable collection in an institution, where it has remained largely intact. It is clear, however, that not all of the Flixton weapons went to Williamsburg. Muskets from the group remained in Neal's own collection and others found their way into other collections. The

The interior of the Powder Magazine, 1975. At this time, many of the Royal Welsh muskets and bayonets were displayed in the bottom tier. The Colonial Williamsburg Foundation.

Royal Armouries, and some of Neal's contemporaries, such as J. C. L. Knapton, Geoffrey Jenkinson and Clay Bedford obtained specimens. Although it has not been proven, it is also thought that a majority of the few brass hilted Royal Welsh Fuzilier infantry hangers known to be in private hands were part of the Flixton Hall collection too. There are certainly other weapons from this collection at large which have not yet been identified as such.

To date, it has not been possible to locate any records of the transaction between W. K. Neal and General Adair, which unfortunately leaves some questions unanswered, the most important of which is just how many weapons formed these displays. At the very least, when Neal purchased and dismantled them, there were 31 muskets, 32 bayonets, 30 or so infantry hangers, 1 spontoon and 7 halberds. However, these arms are an undetermined percentage of the original collection, since numerous others were almost certainly lost in the fire of 1846.

Arms displays are known to have first appeared in English Royal palaces as early as the Restoration, when John Evelyn installed one in Prince Rupert's Guard Chamber at Windsor Castle in 1670.[48] The most famous arms display still in existence, in the Guard Chamber at Hampton Court, was designed and put up by John Harris in 1699. Although many of the weapons are early 19th century replacements, others, like the brass-hilted hangers and plug bayonets, remain from the original installation. Additionally, it is thought that these replacement arms hang in the same configurations as their 17th century predecessors.[49]

The most imaginative display of the era was that in the Small Armoury at the Tower of London, also carried out by John Harris at the very end of the 17th century. Some of the highlights were renditions of "Medusa's Head," "The Back Bones of a Whale," "The Waves of the Sea," and huge columns of pikes topped by Corinthian capitols made of pistols. Most unfortunately, the Small Armoury, and these displays, was destroyed by fire in 1841.

Shortly thereafter, arms displays began to appear in the residences of those prominent individuals who desired and could afford them. In contrast, while these displays were put up in Royal residences as an extravagant display of power, they were installed in private residences as a form of decoration and a show of status. These private displays tended to be more like those at Hampton Court than the Small Armoury at the Tower. Though in Royal palaces these displays were installed in the Guard Chambers, private houses usually didn't contain these rooms, so arms displays were usually put up in either the hall, around the main staircase, or both.

There is one surviving display which is representative of what William Adair's display may have looked like when it was first installed at Flixton Hall in the 18th century. This display graces the walls of Chevening House, Kent, once the seat of Lord Stanhope, and now the official country residence of the British Secretary of State for Foreign Affairs. Like those at Hampton Court, and the one removed from Flixton in 1950, the arms at Chevening are mounted to the wall above the paneled wainscoting, reflecting what was likely a traditional practice. Additionally, the displays at Chevening are centered upon the grand staircase (although not confined to it), just as it was at Flixton Hall.

In 1736, John Harris, famous for his previous work, installed those at Chevening.

The grand staircase and some of the arms displays at Chevening House, Kent, once the seat of the Stanhope family. These displays, installed in 1736, are composed of carbines, pistols, brass-hilted sabers and plug bayonets that were the armaments of a mounted Irish militia unit during the era of Queen Anne's reign. Courtesy of the Trustees of Chevening Estate.

The weapons used had been the armaments of a short-lived mounted Irish militia unit, under the command of Colonel Thomas Pitt, Earl of Londonderry (later Colonel of the 2nd Dragoon Guards, 1715-1726). Since Philip, the 2nd Earl of Stanhope, was Pitt's nephew, the whole lot was purchased for Chevening's adornment for 150 Pounds Sterling. Included are flintlock carbines, pistols, plug bayonets and brass-hilted sabres, all of which date from c. 1700-1715. The vast majority of the firearms are Irish-made, non-Ordnance pieces, but a number of the pistol locks are engraved with the cypher of Queen Anne. Additionally, some of the pistol barrels bear the maker's mark of Robert Wilson.[50] It seems very likely that this gunsmith was related to the James Wilson who is thought to have made the lone 1724-dated "Dublin Castle" musket in the Adair collection (see pages 59-62).

The earliest mention of arms displays at Flixton appeared in the 1840s, as hanging in the great hall.[51] While it seems that the Royal Welsh weapons should have been displayed in the great hall too, if those known today had been hanging there in 1846, then they would have perished in the fire. At that time, these surviving weapons may have been displayed in a part of the house that was undamaged by the fire, like the east wing, or the new offices. Alternatively, they may have been moved from the walls in and around the great hall and grand staircase to protect them from the elements and workmen during the renovation going on at that time.

In 1999, few people were alive who could recall the interior of the mansion, let alone remember the display of arms removed in October of 1950. Lady Darell, General Adair's eldest daughter, had last been at Flixton Hall about the time of the sale, when she was a young lady. Despite the passage of time, in 1999, Lady Darell recalled the weapons displays, but could not describe them in great detail. She remembered that they covered the high walls under the grand staircase and around the triple archways flanking the great hall. The muskets were displayed horizontally in racks, but she was not sure how the swords were arranged. Additionally, "pikes," as Lady Darell described them (perhaps the halberds?), were also part of the displays. Photographs of the areas around the great hall also show close helmets and various flags emblazoned with family crests and arms, supported on spontoon-like staffs.

Fortunately, some professional quality photographs were taken of Flixton Hall's interior by Sir Frederick Adair, circa 1907. While no shot has been located specifically of the weapons displays, one view clearly shows two upright racks of 3 sergeants' halberds against the arch supports between the great hall and the grand staircase. Over one of the arches in a similar photograph, a Royal Welsh Fusilier infantry hanger can clearly be seen crossed with its scabbard on top of dark floral-patterned wallpaper. An earlier photograph of c. 1870 shows four of these halberds and what is possibly one of the infantry hangers hanging on the south wall of the great hall.

A 1999 visit to the remains of Flixton Hall confirmed that a number of walls that formerly held these weapons were still standing, and in reasonably good condition considering the ruinous state of the building in general. Some of the walls even retained large sections of wallpaper. Since the house was emptied, stripped and demolished after the weapons were removed, it can be assumed that any traces found on the remaining walls are evidences of the last items hung there.

In situ, over the very arch adorned with a Royal Welsh Fusilier hanger in the photograph of c. 1907, was found the nail that supported the tip of the blade. This was not an early forged nail, but of the common cut-wire type of the post-1850 period.[52]

Only one of the two walls under the grand staircase survived, and the only trace of its former grandeur was the horizontal beam to which the carved wood paneling was anchored. This beam was attached to the masonry at about 8 feet above the floor. A delineation could also be seen above it that marks where one case of stairs led to the now gone second story.

The grand staircase at Flixton Hall, early 20th century. Note the two racks of three halberds displayed against the arch supports. Some of the weapons displays were on the walls underneath the uprising staircases to the left and right. Courtesy of Lady Darell.

The gutted interior of the west wing where the grand staircase used to stand, 18 February 2000. The blocked-up doorway at the center used to lead to the now vanished drawing room.

Flixton Hall's great hall, as seen from the grand staircase, taken c. 1907 by Sir Frederick Adair. Above the arch at left is one of the Royal Welsh Fuzilier hangers crossed with its scabbard, and in front of the support at center is a rack of three halberds. The helmets along the ceiling moulding appear to be the same ones in the c. 1870 photograph of the great hall (see page 35, top). Courtesy of Lady Darell.

The arch which was crowned with the Royal Welsh hanger and scabbard seen in the previous photograph. The black anomaly above and to the left of the keystone is the dust-covered nail that supported the tip of the sword blade also shown in the previous photograph.

The lofty area formerly nestled between what was an uprising case of stairs and wall paneling, where one group of the Royal Welsh hangers was displayed. Note the two rows of nails, three of which in the top row still retain the brass wire used to secure the hilts.

Situated between these two remains, about 9 to 11 feet above the floor, are two parallel rows of horizontal nails exactly where some of the weapons are known to have once been displayed. These are the traces of what was once one of the sword displays (Neal mentioned two). The rows are about 16 to 18 inches apart and there are 6 to 7 inches between each nail. In fact, three nails in the top row still retain bits of twisted brass wire, suggesting that the hilts of some of the swords were anchored there, with the tips of the blades resting on the bottom row of nails. From a count of the visible nails, it can be assumed that there were at least nine or eighteen swords hanging there, depending on whether or not there was one or two opposing rows of hangers.

On the other side of the great hall, past the other triple archway, rose a smaller, far less elaborate staircase. Only one wall survives here too, but it retains most of its golden-colored floral print wallpaper. It is also situated between the traces of an up-rising case of stairs and an anchor beam for wall paneling. This wall still retains a pair of wooden anchors, set into the plaster about 3 to 4 feet apart horizontally. These are most likely the supports to which a musket rack was attached. Just as Neal remarked about the height at which these weapons were displayed, these traces show that they were between 8 and 12 feet above the floor, mounted over the paneling.

While it may never be known exactly what these displays looked like, Lady Darell's recollections, W. K. Neal's letter to Colonial Williamsburg, the quantity of known components and the evidences on the surviving walls show that the displays were more utilitarian than fanciful. We can be sure that the components of the display were not in the same configuration, or possibly even location, that they were in during the lifetime of William Adair,[53] since Flixton Hall endured a number of major reconstructions and a disastrous fire. The evidence shows that the displays purchased and dismantled by Neal dated, at the earliest, from the remodeling of 1888-92.

A 1952 detail of the grand staircase, showing the same area seen in the previous photograph, shortly before demolition began. Courtesy of the Suffolk Record Office, Ipswich (SPS15046).

Author's sketch of what the arms display on the walls shown in the photos on the bottom of page 56 and on page 55 may have looked like. For the sake of clarity, the dark, patterned wallpaper behind the arms has been omitted.

The wall outside the east side of the great hall, where some of the muskets are believed to have been hung. In the center of the photograph are two wooden blocks set into the plaster which served as anchors to a musket rack.

The Muskets

To the student of British military firearms of the 18th century, the guns in the Adair collection are a highly important assemblage of some extremely rare, and in some cases, unique pieces. To the original collector, William Adair, they were all the same—good examples of the common soldier's service issue musket. They certainly were not rare at the time, and as plain army muskets, they were not at all desirable for a gentleman's shooting needs. Their appeal lay in the fact that Adair was in just the right position to obtain a large quantity of them.[54] His acquisition of these arms may be directly related to his purchase of Flixton Hall in 1753, and his desire to decorate its walls with a theme that reflected his newly acquired "landed gentry" status and his very lucrative livelihood too.

The Dublin Castle musket of 1724

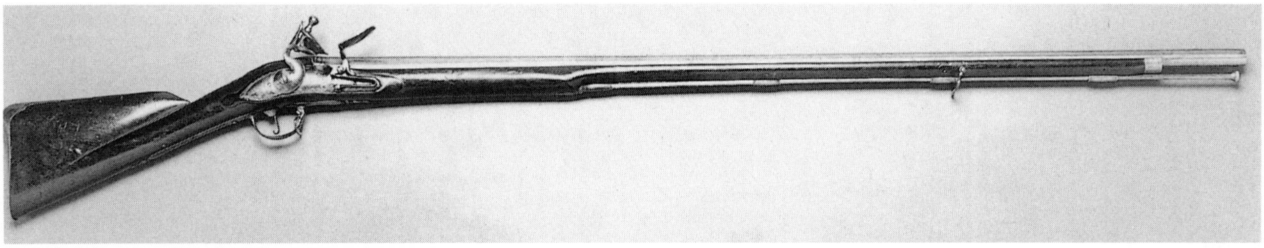

Iron mounted musket for a British regiment serving on the Irish Establishment, c. 1724. This firearm came from the W. Keith Neal Collection, sold at Christie's London, 8 November 1995, as lot 119, and may be a sealed pattern musket. Clinton Miller collection.

The earliest musket yet traced from Flixton Hall is iron mounted, has a 46" barrel, and a bridle-less *1724/Dublin Castle* marked lock, indicating it was made for a unit serving in Ireland. It is the earliest known Dublin Castle musket, and the only Irish-associated musket in the collection. Of additional interest is the possibility that it may be a sealed pattern, due to the remains of two small wax seals on the sideplate-side of the buttstock, very near the buttplate.

Just as the Royal Arsenal supplying Ordnance muskets to units on the British Establishment was the Tower of London, the arsenal at Dublin Castle supplied Ordnance weapons to units serving on the Irish Establishment. The Irish Board of Ordnance operated in much the same way as the British Board of Ordnance, but was completely independent of it. At that time, it was still common practice for both Boards to procure complete weapons, as opposed to procuring just the components and sending them out to be "set up." Unfortunately, a fire at the Irish Public Records Office in 1922 destroyed most of the Ordnance records of that period.[55]

This musket's furniture is typical for iron mounted muskets of the era, and closely similar to that found on the Pattern 1730 musket. Additionally, this musket was originally manufactured with an iron ramrod, almost always the case with iron-mounted muskets of the 1720-1730 era procured for regiments serving on the Irish

The 1724/Dublin Castle marked lock in the lone iron-mounted musket. Clinton Miller collection.

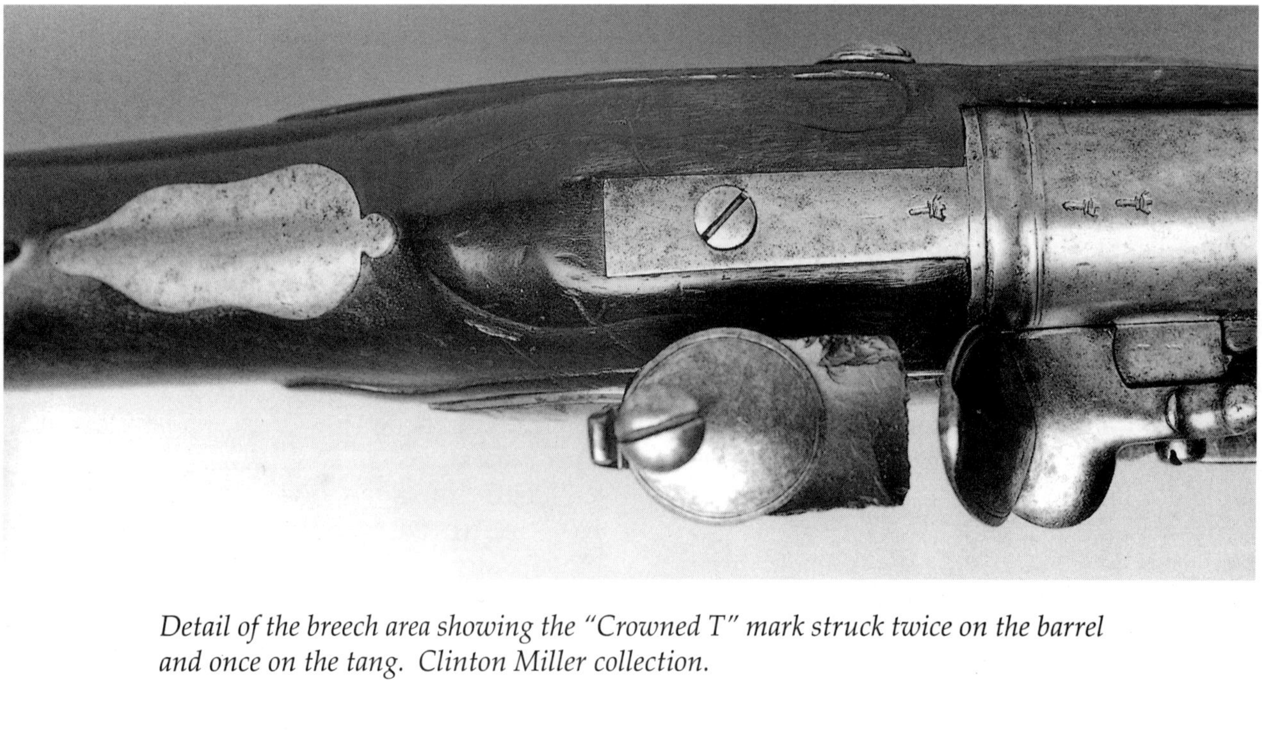

Detail of the breech area showing the "Crowned T" mark struck twice on the barrel and once on the tang. Clinton Miller collection.

The buttplate tang, of the same basic form adopted by the Board of Ordnance for all of its Long Land Pattern muskets, but made of forged iron. The name "WILSON" engraved across the widest part is likely the signature of James Wilson, a Dublin gunsmith known to have been making muskets for British regiments serving in Ireland at the time. Clinton Miller collection.

Establishment. Engraved across the tang of the buttplate is the name WILSON. Normally, this could stand for either the name of the regiment's colonel or, if there are other regimental designations present on the arm, the name of the commander of the company to which this musket belonged. Since there are no other markings present, and there were no regiments at the time commanded by a Colonel Wilson, it seems most probable that this is the signature of the gunmaker who built this weapon.

There was an Ordnance contractor in Dublin named James Wilson who was making muskets for British regiments on the Irish Establishment during this period.[56] His work is also known by an iron mounted 46" barrel musket he made for Hawley's 33rd Regiment[57] which bears the same "Crowned T" mark on the breech of the barrel as the *1724/Dublin Castle* musket. Wilson is also responsible for a stand of similar arms made for Pocock's 8th Regiment.[58] On the muskets for both the 8th and 33rd Regiments, Wilson's name appears engraved on the lockplates, along with company and weapon numbers.

This *1724/Dublin Castle* musket remained in the William Keith Neal collection until November of 1995, when Christie's of London auctioned it in a sale of part of his collection. In the barrel was found a tattered old card tag inscribed by Sir Robert Shafto Adair, 5th Bt., (father of General Sir Allan Adair) which reads;

> Muskets used by the S. Elmham Fencibles (9th troop of Suffolk Yeomanry) raised & commanded by Alexander Adair in 1805, during the Napoleonic Wars. They (used?) the pewter service, part of which (is in?) the hall.

Obviously, the Adair family tradition concerning the weapons collection was erroneous. Early 19th century British yeomanry cavalry did not use obsolete 46" barreled muskets that would then have been between 75 and 50 years old. Since after many generations the real reason for the weapons being at Flixton Hall had been forgotten, they were credited to use by the

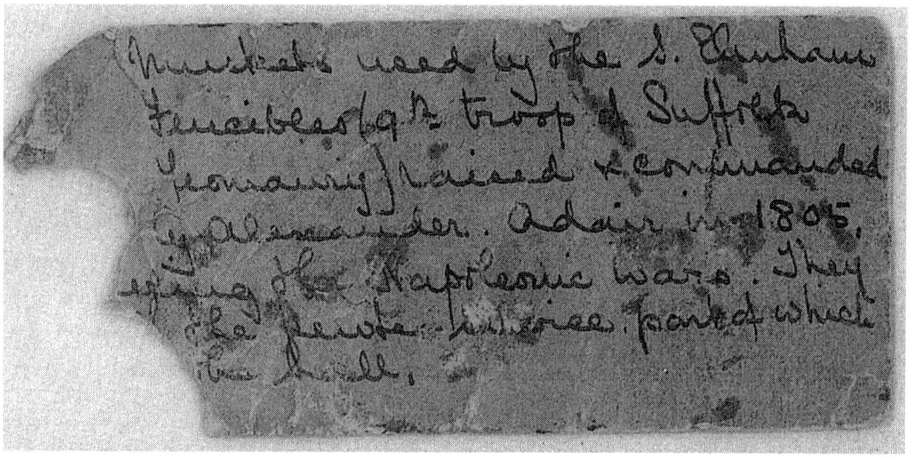

Muskets used by the S. Elmham
Fencibles (9th troop of Suffolk
Yeomanry) raised & commanded
by Alexander. Adair in 1805.
during the Napoleonic wars. They
[illegible] the pewter service, part of which
[illegible] [illegible],

The old tag found inside the barrel of the 1724/Dublin Castle musket. In the hand of Sir Robert Shafto Adair, 5th Bt., it erroneously identifies this musket as having been used by the Suffolk Yeomanry unit raised by Alexander Adair in 1805. Clinton Miller collection.

unit raised by Alexander Adair during the Napoleonic Wars.[59] The aforementioned pewter service used by the officers of this yeomanry unit is still in the possession of the family.

Even though the Adair family had deep roots in Ballymena, William Adair had no ties with the Irish Establishment as an army agent, since British agents did not operate in Ireland, and Irish army agents did not operate in Britain. It is possible that Adair (or Huske?) acquired this musket as obsolete from the stores of a regiment that hired him as an agent after transferring from the Irish Establishment. However, since the Royal Welsh were not in Ireland from 1719 until the end of the century, and they received new arms in 1723 and 1727 made by Freeman of London,[60] this Dublin Castle marked musket was clearly not for them.

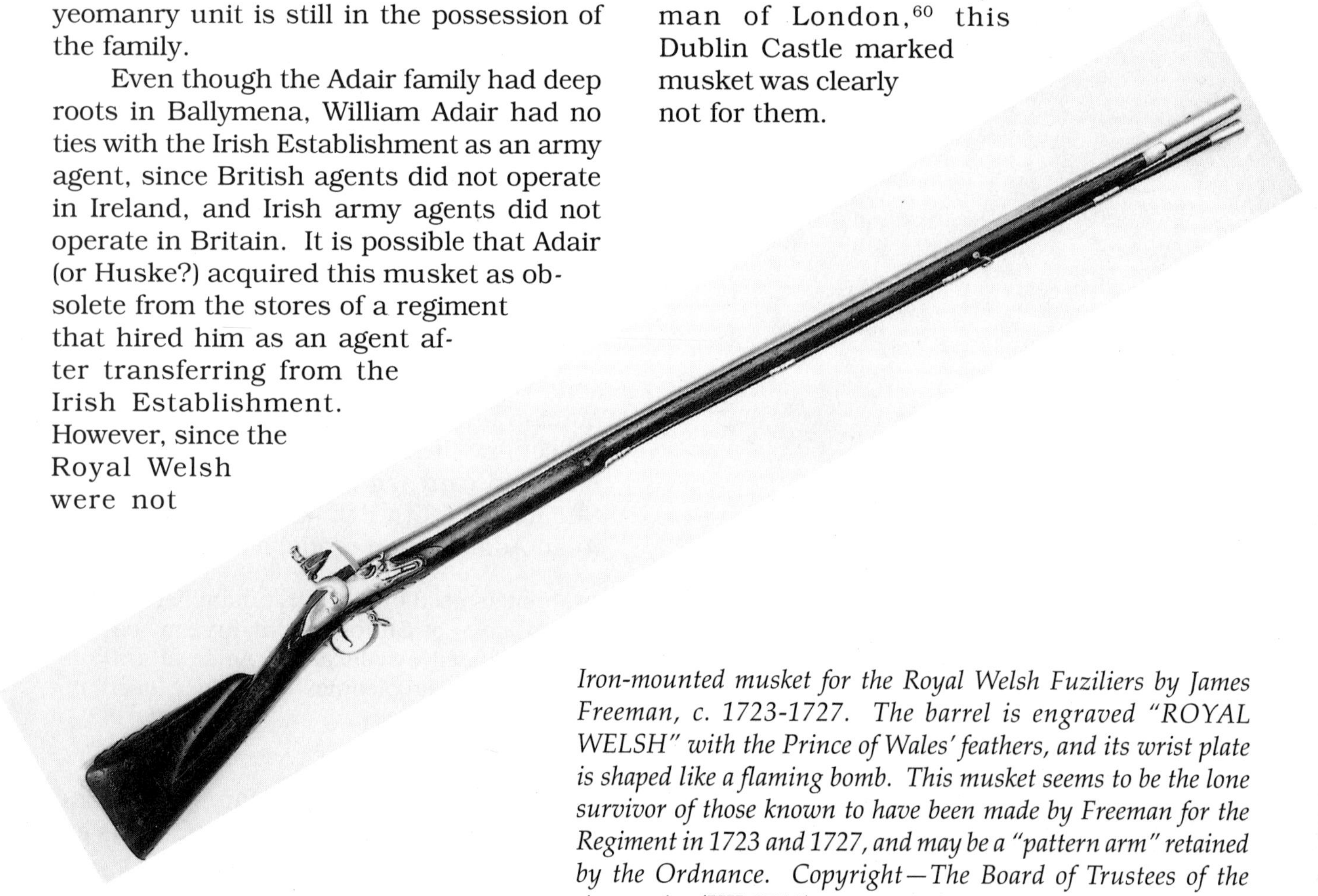

Iron-mounted musket for the Royal Welsh Fuziliers by James Freeman, c. 1723-1727. The barrel is engraved "ROYAL WELSH" with the Prince of Wales' feathers, and its wrist plate is shaped like a flaming bomb. This musket seems to be the lone survivor of those known to have been made by Freeman for the Regiment in 1723 and 1727, and may be a "pattern arm" retained by the Ordnance. Copyright—The Board of Trustees of the Armouries (XII.1749).

The lock of the Freeman musket, also engraved with the Prince of Wales' device. Note the number "14" (in addition to what looks like a 'crowned IG') struck at the breech, which was likely surmounted by a crown. Being that this is a pre-regulation Ordnance-issued arm, this is likely the same inspection mark found struck on the barrels of many Pattern 1730 muskets. Copyright—The Board of Trustees of the Armouries (XII.1749).

The 1739 and 1741 Arms Issues to the Royal Welsh Fuziliers

Right before the outbreak of the War of the Austrian Succession (1740-1748), the Ordnance conducted official inspections of all the muskets and bayonets of the British regiments of foot stationed in Britain. In 1740-1741, due to the deplorable state of many of the arms and the great variety thereof found, the Ordnance, in an effort to achieve standardization, re-armed most of the army with the new muskets it began stockpiling in the 1730s.

This weapon, the "Pattern 1730" Long Land service musket, was the first truly standard common soldier's firearm produced by the British government. At least nine of these muskets were purchased by W. Keith Neal from Flixton Hall in 1950, which accounts for a very significant proportion of these arms that survive in original condition today. All specimens studied bear a sometimes bewildering array of regimental, company and weapon number markings. Of these nine, only one musket is marked to a regiment other than the Royal Welsh Fuziliers.

As will be shown, these muskets were issued under two Royal Warrants, one in 1739 and one in 1741, as a direct result of the Ordnance's efforts to re-arm the British Army with standardized weapons. It should also be taken into account that the Regiment may not have actually received them for quite some time after the date of the warrant, since the wheels of the Ordnance system often moved slowly.

These Pattern 1730s are textbook examples of the type,[61] with the exception of two that have been shortened. Locks are dated between 1727 and 1734, and are marked either *Tower*, *Es. Cole*, *E. Cookes*, or *Iordan*. None of the nine examples have been converted to accept an iron ramrod, and a few still have their original wooden ones. However, all of them have been fitted with a sheet brass noseband (in this case, a unit-applied feature) during their working lives, with the exception of the lone musket marked to Col. Dalway's 39th Regt., which has a modern replaced forestock.

There is considerable variation in the style in which the regimental, company and weapon designations were applied to the Royal Welsh's muskets. What this suggests is that that we are seeing the work of numerous engravers, applying these markings at the company level, possibly at different times. That these markings were not Ordnance-applied can be assumed for a number of reasons. Bailey, in *Board of Ordnance Small Arms Contractors 1689—1840*, lists no engravers employed by the Ordnance during the 1719 to 1743 period, and it was these craftsmen who would have been responsible for the execution of all such markings applied before the arms were delivered to the regiments.

Most importantly, evidence shows that these various types of markings were applied after May of 1742, while the Regiment was campaigning on the Continent, and had been using them for some time. The fact that weapons already in service were being marked, or re-marked, while serving abroad may have something to do with the 1743 change in the Regiment's Colonels, when General John "Daddy" Huske replaced Colonel Peers, killed at Dettingen. Perhaps Huske ordered his Captains to have their companies' weapons marked, if they were not so already. Since these muskets were not manufactured with nosebands, they may have been added at this time too.

1728/TOWER lock in a Pattern 1730 musket marked for Lt. Colonel Waite's Company of the Royal Welsh Fuziliers, c.1743-45. The "crowned broad arrow" struck on the lockplate is of the form usually found on locks dated between 1727 and 1729, but is obscured by the steel spring finial. Note the letter "I" incised into the stock forward of the steel spring. George C. Neumann Collection, Valley Forge National Historic Park.

The engraved marking on the barrel of the Lt. Colonel Waite musket. George C. Neumann Collection, Valley Forge National Historic Park.

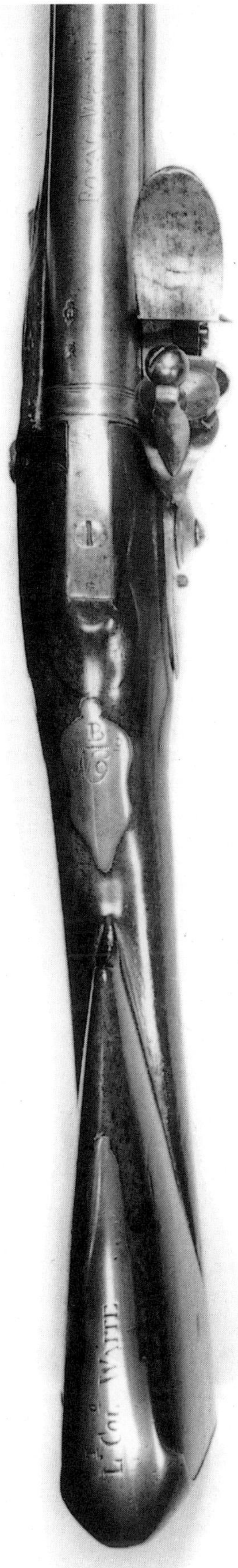

Top of the buttstock and breech area of the Lt. Colonel Waite musket showing the engraved regimental barrel marking, weapon number on the wristplate and the company commander's name on the buttplate. George C. Neumann Collection, Valley Forge National Historic Park.

The variation of markings seen, and the fact that only some of the muskets' stocks were branded with the initials of the commander of the company to which they belonged, further illustrates that these markings were carried out at the company level. This work would have therefore been at the Company Commander's expense, and paid for out of the "contingent," or "non-effective" monies[62] designated specifically for such use.

Six of the musket barrels are marked "ROYAL WELSH" in block letters, while the other two are marked "ROYAL WELSH FUZILIERS" in noticeably smaller block letters. Therefore, these eight muskets can be divided into three markings-based subgroups:

Group I—The "R" and the "W" in ROYAL WELSH will be larger than the other letters. The "rack" number on the wristplate is centered very high and has the "No." engraved in script with the little "o" high and distant to the right. The line separating the company letter from the weapon number is very short, and there is a little dash under the "o." The first letters in the Company Commander's name and rank on the buttplate tang are larger, *en suite* with the engraving on the barrel.

Group II—All of the letters in ROYAL WELSH are of equal height, and are slightly smaller than those of Group I. The weapon number is centered low on the wristplate and the company letter is set very high. The "No." is engraved in stiff, thin block letters. All of the letters in the Company Commander's name and rank engraved on the buttplate tang are of uniform size (with the exception of the tiny letter "T" used in the abbreviation of "Captain"). All three specimens have the initials of the Company Commander branded into the stock, and one musket has no alphabetical company designation engraved on the wristplate.

Group III—All of the letters in ROYAL WELSH FUZILIERS are of equal height and are much smaller than those in the first two groups. The "rack" number is well centered on the wristplate with considerable distance between the company letter and weapon number. The line separating them bisects the wristplate and the "No." is executed in a script different to that on Group I's wristplates. The weapon number is also engraved on the buttplate tang (but rotated 180 degrees) along with the re-engraved Captain's name *BERNARD*. Both recorded examples have the initials of John Bernard branded into the buttstock.

Pattern 1730 RWF Musket Markings

Lockplate	*Barrel*	*Wristplate*	*Buttplate*	*Stock*
Group I				
1728/Tower	Royal Welsh	B/No.9	Lt. Col. Waite	—
1734/Iordan	Royal Welsh	C/No.2	Maj. Lort	—
1727/Tower	Royal Welsh	G/No.62	Capt. Cary	—
Group II				
1729/Cookes	Royal Welsh	No.19	Capt. Taylor	AT
1741/Tower	Royal Welsh	H/No.16	Capt. Hickman	WH
(replaced)	Royal Welsh	H/No.18	Capt. Hickman	WH
Group III				
1730/Cole	Royal Welsh Fuziliers	H/No.39	Capt. Bernard No.39	IB
1734/Iordan	Royal Welsh Fuziliers	H/No.4	Capt. Bernard No.4	IB

Conveniently, all eight Royal Welsh muskets have the surname and rank of the Company Commander engraved on the buttplate tang, which is vital in determining the time frame in which these muskets were marked and used. As listed by Kirby in *Officers of the Royal Welch Fusiliers (23rd Regiment of Foot), 16 March 1689 to 4 August 1914*, these Company Commanders, and the dates served in their respective ranks are:

Lt. Col. John Waite
Commissioned Lt. Col. on 11 April 1743 and served in this rank until 15 February 1747. However, Waite continued to serve as a Captain of a Company until 4 June 1748.

Waite was first commissioned as an Ensign in the 14th Regiment in 1718, and entered the Royal Welsh as a Captain on 25 December 1722. He was promoted to Major on 4 September 1739, and toward the end of his career in the Regiment, was appointed Lt. Governor of Pendennis Castle.

Maj. Roger Lort
Commissioned Major on 18 April 1743 and killed at the Battle of Fontenoy, 11 May 1745.

Lort received his first commission as a Lieutenant on 11 April 1708, and joined the Regiment as a Captain on 16 July 1730. At St. Mary's Church, Tenby, there is a monument to Lort, commemorating both his military service and his personal virtue.

Capt. John Bernard
Commissioned Captain on 8 April 1742 and promoted to Major 22 February 1748.

Bernard was born in Ireland in 1702, and was first commissioned into the Regiment on 13 March 1719 as a 2nd. Lieutenant, in place of his brother William, who had been promoted that same day. He was next promoted to Lieutenant on 25 November 1731. Captain Bernard was wounded while leading his Company at the Battle of Fontenoy on 11 May 1745, and died on 18 June 1755, at the age of 53.

John Bernard's son, Benjamin Bernard, was commissioned into the Regiment in 1748, and by 1771, was Lt. Col. of the Royal Welsh Fuziliers. Lt. Col. Bernard took the Regiment to America in 1773, and was wounded in the thigh on 19 April 1775, at the Battles of Lexington & Concord, while leading the Royal Welsh's battalion companies as part of Percy's relief column.

Capt. James Cary
Commissioned Captain 10 August 1737 and retired 10 January 1749.

Cary was first commissioned into the Regiment as a 2nd. Lieutenant on 1 May 1710, and made Lieutenant late the next year. In June of 1715 he is listed as serving in Capt. Jones' Grenadier Company, and by 1723 was recorded as serving in Lt. Col. Peers' Company. By 1727 he was serving in Capt. Bissell's Company.

Capt. William Hickman
Commissioned Captain 23 March 1731 and promoted to Major 27 May 1745, to replace Maj. Lort, killed at Fontenoy.

Hickman was born in 1696, and may have been the younger brother of Henry Hickman, who served as a Royal Welsh Fuzilier officer from 1710-1742. William first entered the army as an Ensign in Kirke's 2nd Regiment of Foot on 1 July 1717. On 25 December 1722 he entered the Royal Welsh as Captain Lieutenant. Captain Hickman was recorded as having been wounded at the Battle of Fontenoy. He died in the parish of St. Martin's-in-the-Fields on 8 January 1756.

Capt. Arthur Taylor
Commissioned Captain 25 April 1741 and retired 30 May 1747.

Taylor entered the army as a Lieutenant in the 35th Regiment in 1713. He was placed on half-pay before transferring to the Royal Welsh on 11 August 1718, to serve as a Lieutenant in Capt. Pigott's Company. Captain Taylor was posted as missing after the Battle of Fontenoy on 11 May 1745, but was only wounded, and survived to retire from the Regiment two years later.

Based on the dates that these men served at the ranks engraved on the buttplate tangs, we can initially deduce that these muskets were issued and marked for the Regiment between 1731 and 1749, a parameter which can be greatly narrowed. In order to do this, the history of the Royal Welsh and its regimental structure must be examined, in conjunction with the evidence provided by the weapons themselves, and

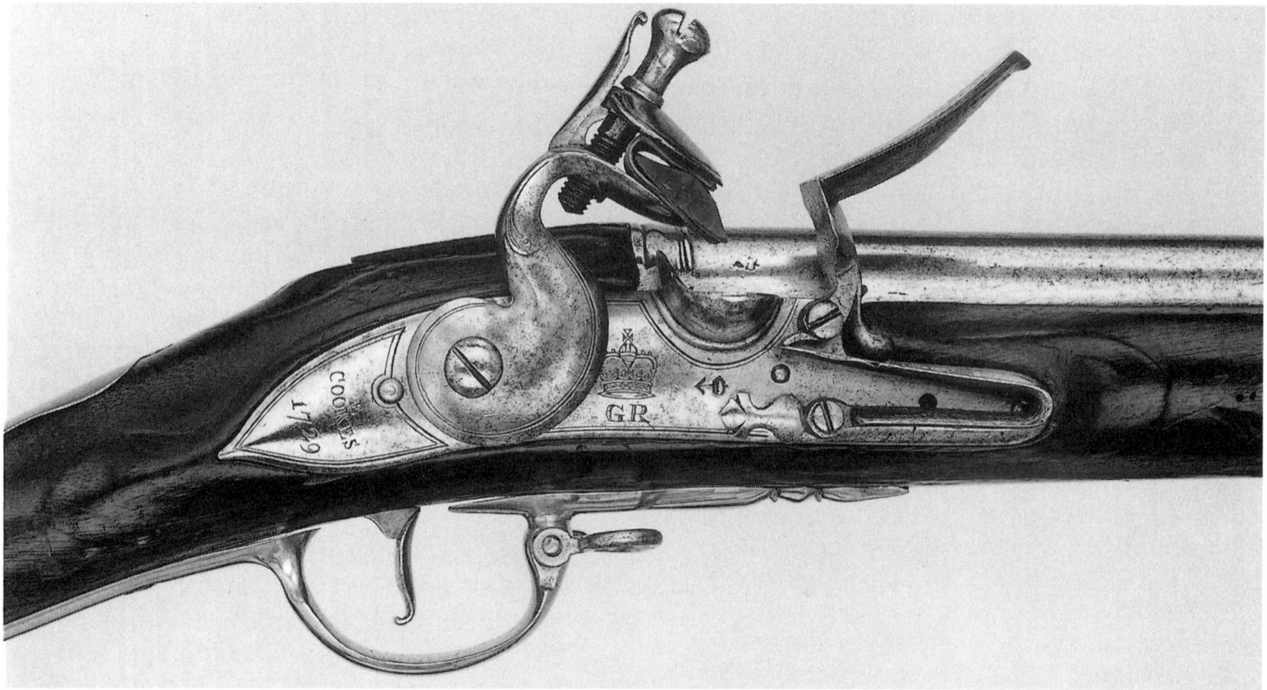

1729/COOKES lock in a Pattern 1730 musket issued to Capt. Taylor's Company of the Royal Welsh Fuziliers. This lockplate has the same "crowned broad arrow" found on the 1728/Tower lock in the Lt. Colonel Waite musket, also seen on a few Pattern 1727 bayonets (see pages 99 and 100). A partially struck "Crowned 14" inspector's number is visible on the breech above the vent. The Colonial Williamsburg Foundation.

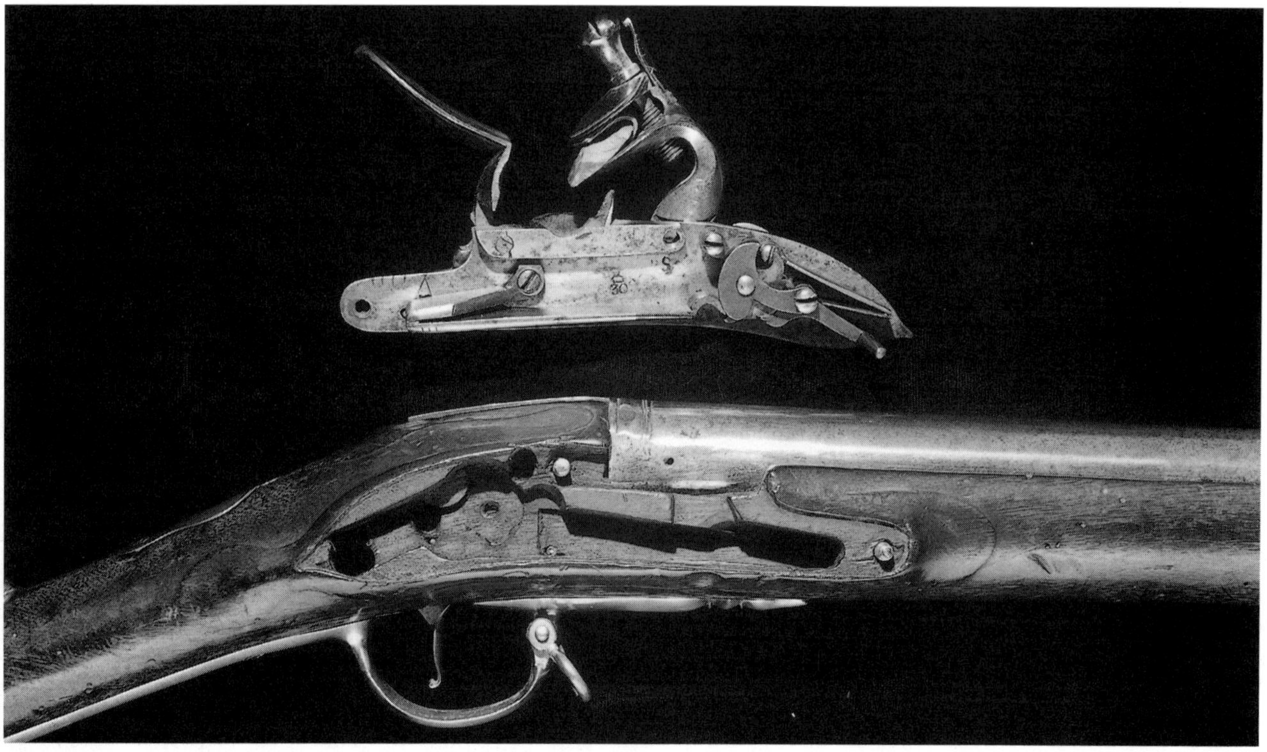

The lock mortise and mechanism of the Captain Taylor musket. Note the various inspectors' and makers' marks and how well the inletting corresponds to the rear of the lockplate. The Colonial Williamsburg Foundation.

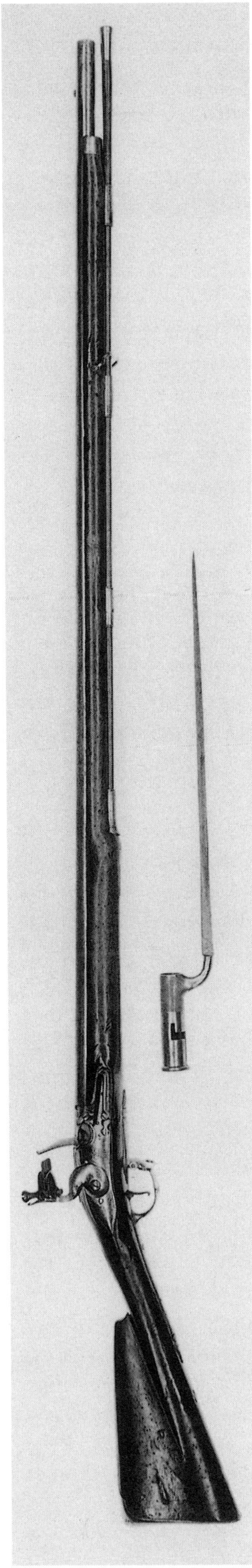

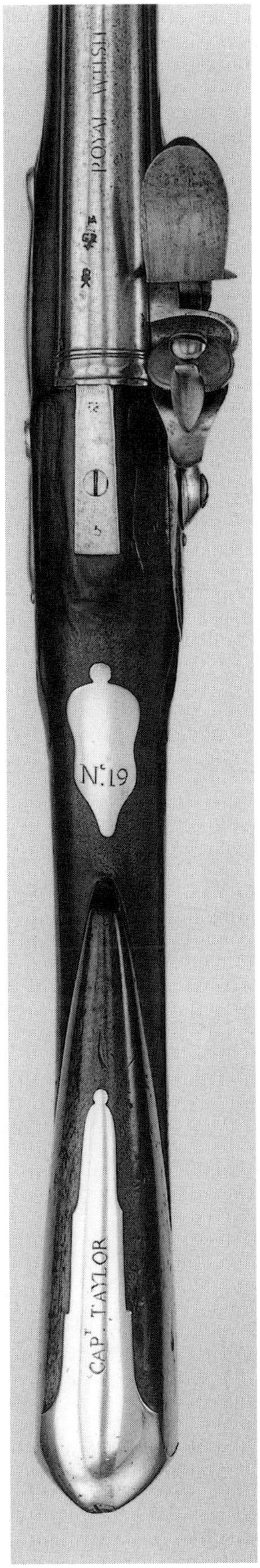

(Far left) Overall view of the Captain Taylor musket. The bayonet pictured below it is incorrect for this musket. The Colonial Williamsburg Foundation.

(Left) Top of the buttstock and breech area of the Captain Taylor musket showing the engraved barrel, wristplate and buttplate markings. For an unknown reason, this musket's wristplate was never given a company letter designation, making it unique in the Adair Collection. The Colonial Williamsburg Foundation.

The initials of Captain Arthur Taylor branded into the stock behind the trigger guard of the Pattern 1730 musket issued to his company. Three other Pattern 1730s from the Adair collection have the initials of the company commander to whom they were issued branded into the stock. The Colonial Williamsburg Foundation.

the warrants issuing them. From Cary & McCance's *Regimental Records of the Royal Welch Fusiliers, Vol. I*, during the 1731 to 1749 period (new-style dates), a summary of the Regiment's more important events includes;

1731-1738 – various garrison and policing duties in England and Scotland

1739

24 October – death of General Joseph Sabine, Colonel of the Regiment

23 November – Lt. Col. Newsham Peers promoted to Colonel of the Regiment

1742

May – departed England for the Low Countries

1743

16 June – Battle of Dettingen

" " – death of Colonel Newsham Peers, from wounds received

28 July – General John "Daddy" Huske appointed Colonel of the Regiment

1744

September – two additional companies formed in England, for territorial duty

1745

11 May – Battle of Fontenoy (defeat)

9 July – taken prisoner at Ghent

October – landed back in England

1747

28 January – departed England for Low Countries

21 June – Battle of Lauffeld (defeat)

1748

7 October – Treaty of Aix-la-Chapelle ends the war

8 November – two additional companies formed in 1744 disbanded

December – landed back in Scotland

According to War Office records at the Public Records Office, London (PRO), during the 1731 to 1749 period, the Regiment was issued with new weapons by Royal Warrants of the following dates;

2 July 1739 – 110 muskets and bayonets for the 11 men added to each of the 10 companies.

9 July 1741 – a complete new set of 620 muskets and bayonets to replace the old worn-out ones.

23 July 1744 – 146 muskets and bayonets for the 2 new augmented companies of 70 privates and 3 corporals.

21 Nov. 1745 – a complete new set of 726 muskets and bayonets to replace those lost at the taking of Ghent.

The 1739 issue provided eleven additional muskets and bayonets to the same number of men added onto each of the ten companies, thereby established at seventy Privates and three Corporals. It can be assumed that these 110 firearms were Pattern 1730s, and were likely the first brass-mounted muskets the Regiment received. Previous to this date, the Regiment had been armed with iron-mounted muskets made by James Freeman of London, received in 1723 and 1727. A superb example of one of the muskets made by Freeman for the Regiment survives in the collections of the Royal Armouries (see note 60 and photos on pages 62 and 63).

As a result of the Ordnance's inspection of the Regiment's arms in 1739 or 1740, it was deemed necessary to replace those of the whole regiment, excepting the eleven new arms provided to each of the ten companies in 1739. Therefore, a Royal Warrant dated 9 July 1741 provided for sixty-two muskets and bayonets to be issued to each of the Regiment's ten companies.

The July 1744 issue can be ruled out as having supplied any of the muskets from Flixton Hall. All ten Companies of the Regiment were serving in Flanders when the order was given to raise two additional companies of seventy privates and three Corporals for service in Britain. The 1744 issue was intended to arm these two new companies, which were commanded by Captains Matthew Fortescue and William Hepburne. These 11th and 12th Companies of the Regiment primarily marched around Britain guarding against invasion, and were disbanded in 1748. Since neither Fortescue's nor Hepburne's name appears on any of the buttplate tangs, it can be confidently stated that none of the studied Royal Welsh Fusilier Pattern 1730 muskets were of this issue.

The warrant of 21 November 1745 can also be ruled out as having been responsible for the issue of any of the Regiment's weapons in discussion. After the Battle of

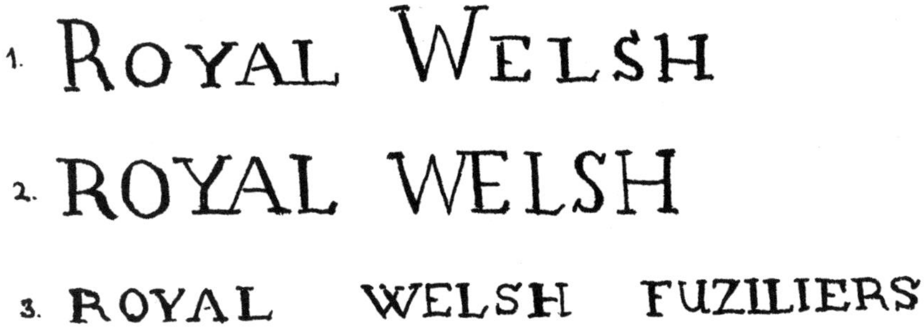

A comparison of the three different types of regimental markings found on the Pattern 1730 muskets.

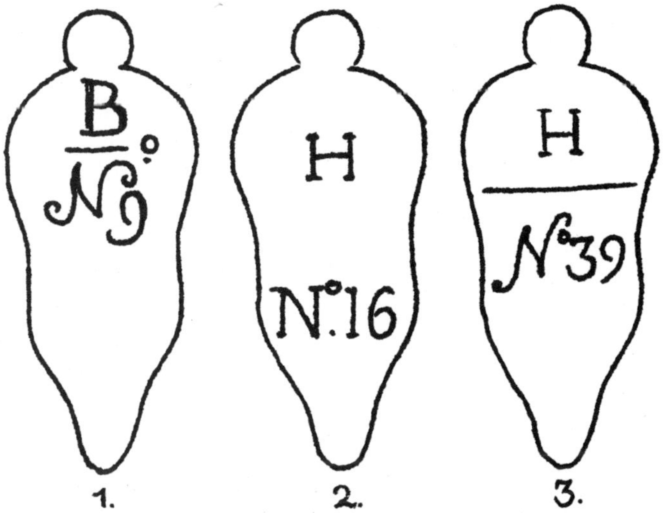

The three different types of company and weapon number markings on the wrist plates of the Pattern 1730 Royal Welsh muskets.

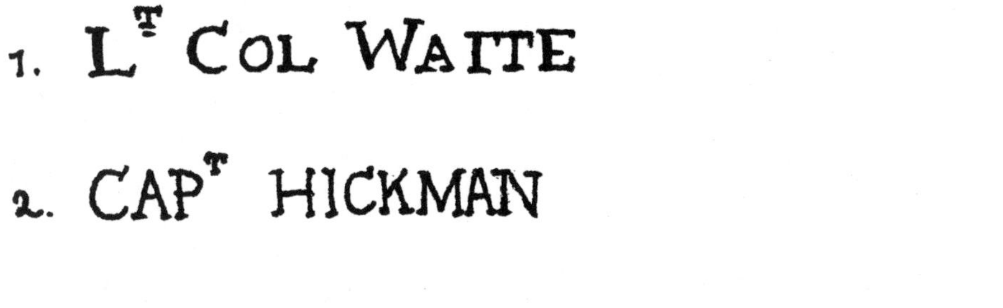

The Company Commander's name and rank as engraved on buttplate tangs of the three types of Pattern 1730 Royal Welsh muskets.

purſued them till 7 a Clock, when the King rode over the Field of Battle. We are almoſt ſtarved, I have had nothing but Ammunition Bread this Fortnight, and Water or ſour Wine. Before we engaged we threw all our Baggage away, ſo that we are obliged to lie with our Horſes in our Hands in the open Field. We fought the Battle in a Field of Rye, our Horſes have nothing but green Corn this Fortnight, of which we have deſtroy'd thouſands of Acres. Tho' we far'd ſo hard our Courage did not fail us.

In the Camp near Hanau, June 10. *From an Officer in Col.* Piers'*s Foot, the* Welſh *Fuſiliers.*

THE *Meanders* of the *Mayne* favour'd the Enemy ſo much, that their Cannon fronted, flanked, and rear'd us, and did great Damage. Our Men were eager to come to Action, and did not at all like the long Bullets (as they term'd 'em) for indeed they ſwept off Ranks and Files. However, when we came to the ſmall ones, they held them in ſuch Contempt, that they really kept the ſame Order as at any other Time. Their Gens d'Armes broke thro' ſome of our Regiments Sword in Hand; but very few of them returned: However, General *Bland*'s Regiment ſuffer'd greatly, having loſt one half, I really believe. *Philip Honywood*(their Major) received many Wounds, and lay 6 Hours (quite ſtripp'd) in the Field of Battle: There is now Hopes of his Recovery. If the Blue Guards, and ſome others of our Horſe and Dragoons had done their Duty as he did, few of the *French* had eſcaped. We were obliged to make an Interval to let the Blues through us. Brigadier *Huske*, who was at the Head of our Regiment, exhorted them, in the beſt Manner that could be, to return to their Charge, but to no Purpoſe; and we were forc'd to let them paſs, for our own Security, otherwiſe we had been broke to Pieces by them. The only Huzza the *French* gave, was at their Retreat, and that but a feint one. Our Army gave ſuch Shouts before we were engaged, when we were about 100 Paces apart before the Action began, that, we hear by Deſerters, it brought a Pannick amongſt them. We attack'd the Regiment of *Navarre*, one of their prime Regiments. Our People imitated their Predeceſſors in the laſt War gloriouſly, marching in cloſe Order, as firm as a Wall, and did not fire till we came within 60 Paces, and ſtill kept advancing; ſo that we had ſoon cloſed with the Enemy, if they had not retreated: For when the Smoak blew off a little, inſtead of being among their Living, we found the Dead in Heaps by us; and the ſecond Fire turn'd them to the Right about, and upon a long Trot. We engaged two other Regiments afterwards, one after the other, who ſtood but one Fire each; and their Blue *French* Foot Guards made the beſt of their Way without firing a Shot. Our Colonel fell in the firſt Attack, ſhot in the Mouth, and out at the Neck; but there are Hopes of his Recovery. The Gens d'Armes are quite ruin'd, who are their chief Dependance, and intended to cut us all to Pieces without firing a Shot. Our Regiment ſuſtain'd little Loſs, tho' much engaged; and indeed our whole Army gives us great Honour. Brigadier *Huske*, who behaved gloriouſly, and quite cool, was ſhot thro' the Foot at the Time that our Colonel fell, yet continued his Poſt. We have no more than 50 kill'd and wounded, and one Officer beſides the Colonel. What preſerved us, was our keeping cloſe Order, and advancing near the Enemy ere we fir'd. Several that popp'd at 100 Paces loſt more of their Men, and did leſs Execution; for the *French* will ſtand Fire at a diſtance, tho' 'tis plain they cannot look Men in the Face. 'Till we obtained this Victory we wanted Bread, and 'tis not to be imagin'd what Fatigues the Army underwent by continual Alarms. 'Till now, I aſſure you, I have not been under Cover above two Nights in 14. The Night after the Action, it rain'd without Intermiſſion 'till 8 next Morning, and very violent. *T. R.*

P. S. *Cornwallis's*, the *Scotch* Fuſileers, and *Johnſton's* Regiments were all diſorder'd by the Gens d'Armes, who had all Cuiraſſes, *viz.* Breaſt-Plates and Back-Plates down to the Waiſt, which were in great Numbers diſpers'd in the Field: I would have ſecur'd a Set of them, for Curioſity ſake, but they were too weighty to carry off.

The Blues faught bravely at laſt.

From the Camp of the Allies at Hanau.

A General Thankſgiving was religiouſly performed in the whole Army on the 24th of *June*. On *July* 2, the *French* Army which had kept over againſt us, on the other Side the River ſince the Battle, marched very early in the Morning in a precipitate Manner towards *Oppenheim*, on which a Party was ſent after them, who took ſome Baggage and brought in Priſoner the Pr. of *Craon's* Son who lay wounded at a Village. On the 8th an *Auſtrian* Captain, with 50 Foot and 30 Dra-

An account of the Battle of Dettingen, from the July 1743 issue of The Gentleman's Magazine. *Penned by an unnamed officer of "Col. Pier's Foot, the Welsh Fusiliers," it is signed "T.R." According to Kirby, the only officer in the Regiment at the time who had these initials was Lieutenant Thomas Rodd, who retired after being wounded at Fontenoy in 1745. Perhaps Lt. Rodd served in a company to which one of the Pattern 1730 muskets belonged?*

1730/Es. Cole lock in a Pattern 1730 musket re-issued to Captain Bernard's Company of the Royal Welsh Fuziliers, c. 1742-45. The "crowned broad arrow" struck on the lockplate is of the type normally seen on Ordnance lockplates dated between 1730 and 1738. Private collection.

Breech detail of the Captain Bernard musket No. 39, illustrating the somewhat worn regimental markings and the Ordnance proof marks. Private collection.

The initials of Capt. John Bernard branded into the buttstock of musket No. 39 issued to his company of Royal Welsh Fuziliers. Private collection.

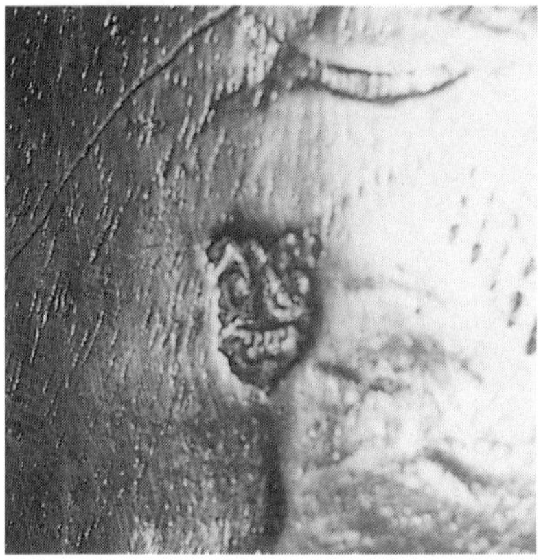

Ordnance storekeepers mark of a crowned Royal Cypher stamped into the lock-side of the buttstock of Capt. Bernard musket No. 39. The top of the crown points toward the buttplate. Private collection.

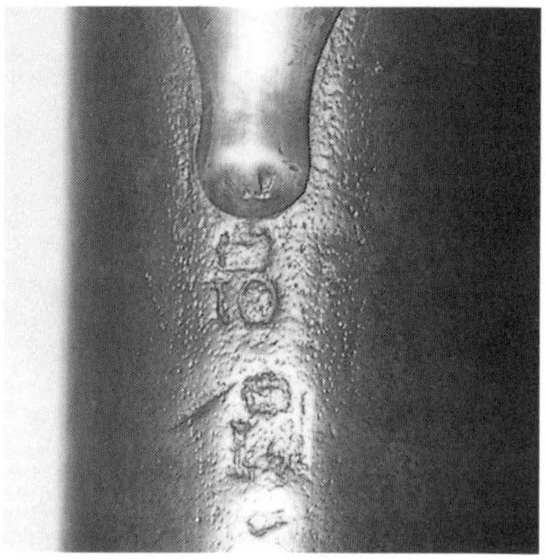

Ordnance inspector's marks of a "crowned 10" and a "crowned 14" stamped into the stock of Capt. Bernard musket No. 39 at the tail of the trigger guard. Private collection.

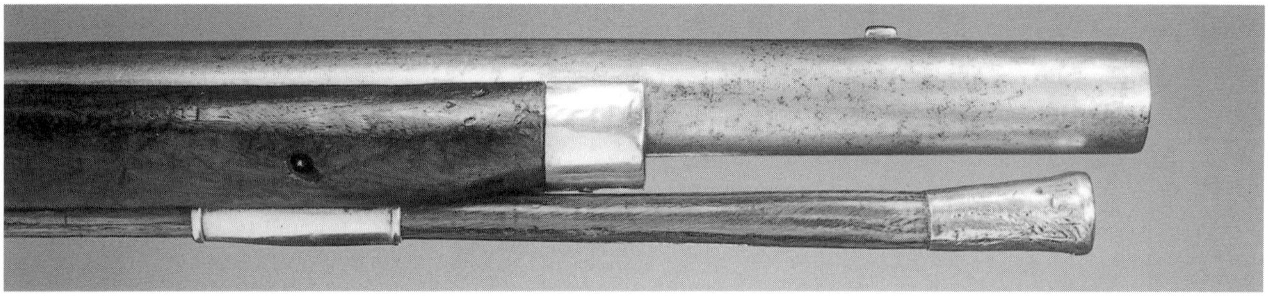

The muzzle area of the Captain Bernard musket No. 39, showing the sheet brass noseband and the sheet brass tip on its original wooden ramrod. Private collection.

Fontenoy, the Royal Welsh were successfully besieged at Ghent and lost all of their arms, ironically surrendering them exactly four years after the date of the Royal Warrant which issued most of them. Once exchanged, the Royal Welsh arrived back in England in late October of 1745, and were re-armed again with a complete new set of 726 muskets and bayonets from the Board of Ordnance (no muskets from this issue have yet been identified).

Based on the evidence discussed, it can be stated that these eight muskets could only have been supplied to the Regiment by the Royal Warrants of 2 July 1739 and 9 July 1741. However, each group must be looked at individually, since these markings vary greatly, and were apparently applied at the Company level, and may not have all been done at the same time.

Group I—Based on the time of Roger Lort's service as Major of the Regiment, this group can be dated as having been marked during the April 1743 to May 1745 period. Since the Regiment was actively serving on the Continent at this time, these designations were certainly applied there.

Group II—Since William Hickman was promoted to Major in May of 1745, and Arthur Taylor was promoted to Captain in April 1741, it can be stated that these muskets were marked during that span. Confusingly, two muskets from this group and both muskets from Group III are marked to "H" Company, but from the commands of two

Lieutenant Joseph Sabine, Royal Welsh Fuziliers, c. 1740, by William Hogarth. Sabine, first commissioned into the Regiment on 17 July 1739, was the son of General Joseph Sabine (by his second wife) Colonel of the Royal Welsh from 1705 until his death in 1739. Lt. Sabine's uniform is identical to that worn by his cousin, Colonel Newsham Peers (see page 17), with the addition of a gilt gorget suspended from a royal blue ribbon. Sabine was promoted to Captain on 22 April 1743 (in place of his brother John) and was posted as missing after the Battle of Fontenoy. On 29 May 1745 he was listed as "dead" at Lessines. This portrait (current location unknown) was reproduced in color and discussed by The Rev. Percy Sumner in the Summer, 1944 issue of the JSAHR (Vol. XXII No.90) on pp. 219-220. Photo by the National Portrait Gallery, London.

different Captains (William Hickman and John Bernard) but it has not yet been possible to determine which Captain preceded the other in command of "H" Company.

Group III—This group has proven to deviate from the first two for a number of reasons. Firstly, they were obviously re-designated during their time of service within the Royal Welsh. On these two weapons, the same hand engraved the "*Capt.*" on the buttplate tangs and the name of the Regiment on the barrels. However, the Captain's surname "*BERNARD*" has been engraved in a different, heavier hand. Since there are also traces of file marks in these areas and vise marks in the buttstocks of both weapons, it would seem that these two Pattern 1730s were originally issued to a Captain other than John Bernard, and subsequently altered once Bernard took over "H" Company. It should also be noted that the engraver who altered the buttplates of Group III to Capt. "*BERNARD*" appears to have marked the arms of Group I too.

Since Captain Henry Hickman (older brother of Capt. William Hickman?) retired from the regiment on 8 April 1742, the same day John Bernard was promoted to Captain, it may well be that John Bernard purchased Henry Hickman's commission and took over his Company (an activity that would have required the assistance of William Adair).

All of this shows that the Group III muskets were marked previous to April 1742, when Bernard took over the company, and may be the earliest group to have received their markings. Assuming this to be true, then in May of 1742, the Group III muskets left England for the Continent already engraved (unlike those of Group I, and possibly Group II).

An interesting period reference to a musket from Captain Cary's Company, certainly emanating from the November 1745 issue, should be noted. This letter is important in that it shows that the post-Ghent issue of muskets was marked in a similar fashion to those issued in 1739 and 1741. During the Jacobite Rebellion, on 2 April 1746, the Secretary at War wrote to the Commanding Officer of the Royal troops at Carlisle about a letter he had received from the Mayor of Kendal concerning stores captured from Bonnie Prince Charlie's rebels. Amongst the captured arms and gunpowder was:

> One firelock with slings marked on the barrell 'Royal Welsh' on butt 'Captain Cary 9/58.'

It is possible that the company letter "G" was mistaken for a "9," in which case this firearm was only 4 numbers away from the 1739 or 1741 issued Capt. Cary G/No.62 musket from Flixton Hall. However, should this account be correct, then the markings applied to the muskets from the November 1745 issue had numerical company designations in place of alphabetical ones, and were therefore different from those applied to the arms issued in 1739 and 1741. Given that the Regiment was never within 200 miles of the line of march of the rebels, it may be that this musket was disposed of by a deserter from Capt. Cary's Company.

Unlike the later models issued to the Regiment in 1756 and 1757, these Pattern 1730s saw active service, and show varying degrees of wear and evidence of moderate to hard use. Most have deep vise marks impressed into their stocks (left by the engraver?) and visible wear to the engraved markings. Their buttplates bear the innumerable contact marks of hitting the ground hundreds, if not thousands of times during the "manual exercise." Two muskets even bear the personalized graffiti of private soldiers. The Capt. Bernard No. 39 musket has the initials "GM" carved into the left side of the buttstock, and the Lt. Col. Waite musket has the letter "I" incised into the stock forward of the steel spring. Should a roster from Captain Bernard's Company survive, the identity of "GM" might be revealed.

The musket now in the collection of The Royal Welch Fusiliers Museum (#3026-A) has undergone a number of interesting al-

The Major Lort musket

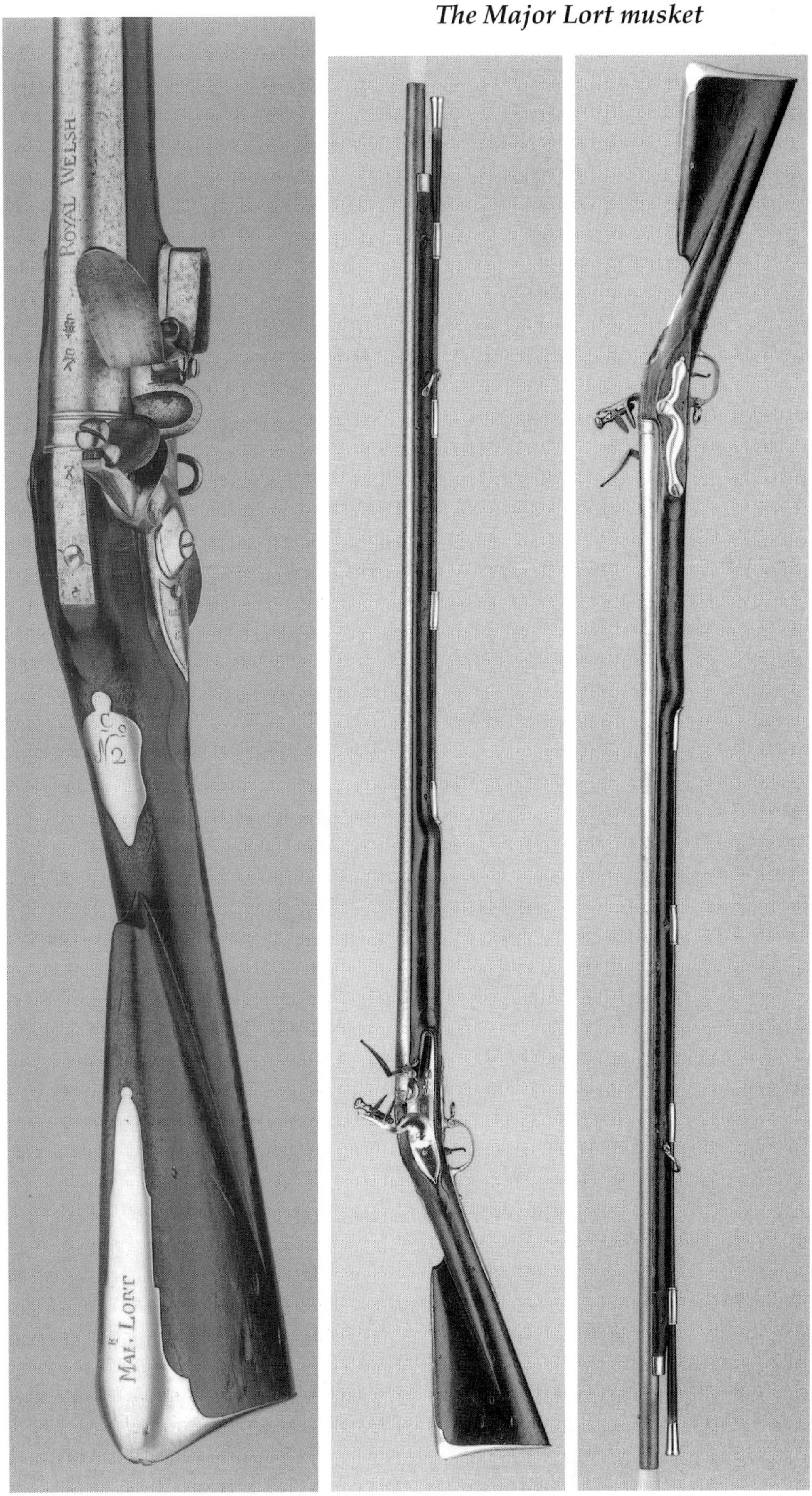

(Left) Top of the buttstock and breech area, showing the barrel, wristplate and buttplate markings.

(Center) Overall view of the musket from the lock-side, exemplifying the Pattern 1730 musket in its classic form. The sheet brass noseband is a post-manufacture addition, probably executed by a regimental armourer.

(Right) Overall view of the musket viewed from the sideplate side.

The Colonial Williamsburg Foundation.

The swell of the forestock at the tailpipe of a Pattern 1730 musket. On later muskets, the size of this swell was greatly reduced. Private collection.

Forends of two Pattern 1730 muskets showing the sheet brass nosebands added after their original manufacture. The Colonial Williamsburg Foundation.

terations.[63] It has received a second lock of a slightly later date and type, engraved *WOOD/1748* at the tail. This can be confirmed since it was issued to "Capt". Hickman's Company, and Hickman was promoted to Major three years before this lock was made. Additionally, a 1748-dated lock would not have been originally fitted to a Pattern 1730 musket.[64]

This musket has also had the last few inches of its forestock replaced at an undetermined point, probably when its barrel was shortened to 44 inches. The bayonet lug, upper ramrod pipe and sheet brass noseband were properly relocated. The Capt. Bernard musket No. 4 (with a *1734/ Iordan* marked lock), has also had its barrel cut down at a later date, but in this case to 42 inches. Its stock was also appropriately shortened, and its hardware adjusted accordingly. Interestingly, its barrel is vented so far from the center of the pan that part of the latter had to be ground away to allow the flash from the priming to reach the main charge in the breech.

In the lineage of the Long Land Pattern musket, the next type is only a transitional, hybrid variant. The Pattern 1730/40 is represented by two muskets from Flixton Hall,

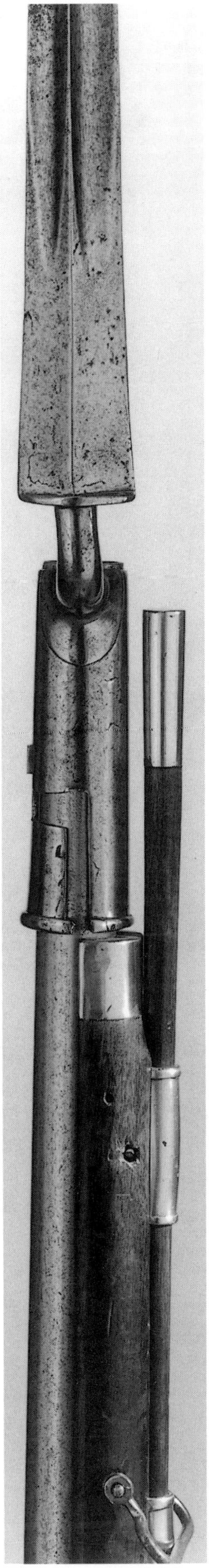

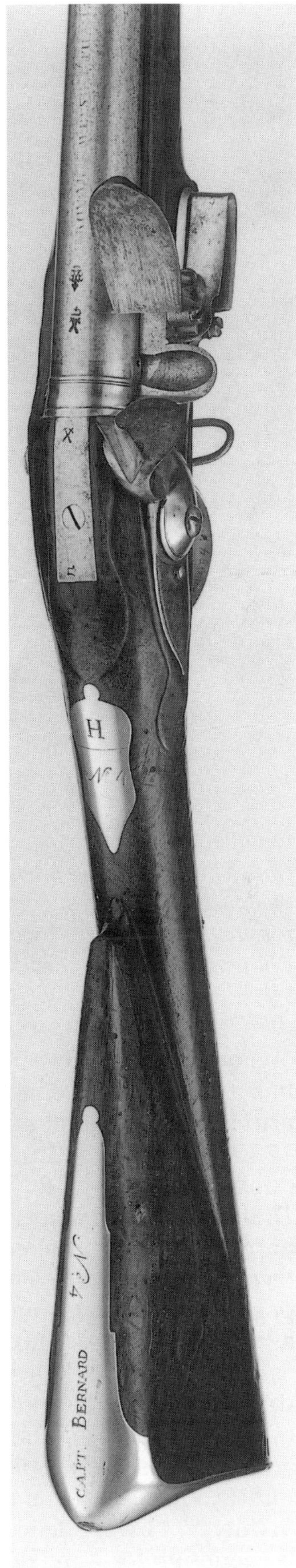

Top of the buttstock and breech area of Pattern 1730 musket number "4" issued to Capt. Bernard's Company of the Royal Welsh Fuziliers, c. 1742-45. Note the duplicate weapon number engraved on the buttplate tang and the addition of the word "FUZILIERS" to the more common "ROYAL WELSH" barrel engraving. The Colonial Williamsburg Foundation.

Muzzle of Captain Bernard musket No. 4, mounted with a bayonet. This is one of two Pattern 1730 muskets from Flixton Hall that have been shortened at the muzzle, in this case by 4 inches. Note how close the top ramrod pipe is to the second one. The bayonet lug, one barrel-pin loop and the noseband have all been relocated appropriately. The Colonial Williamsburg Foundation.

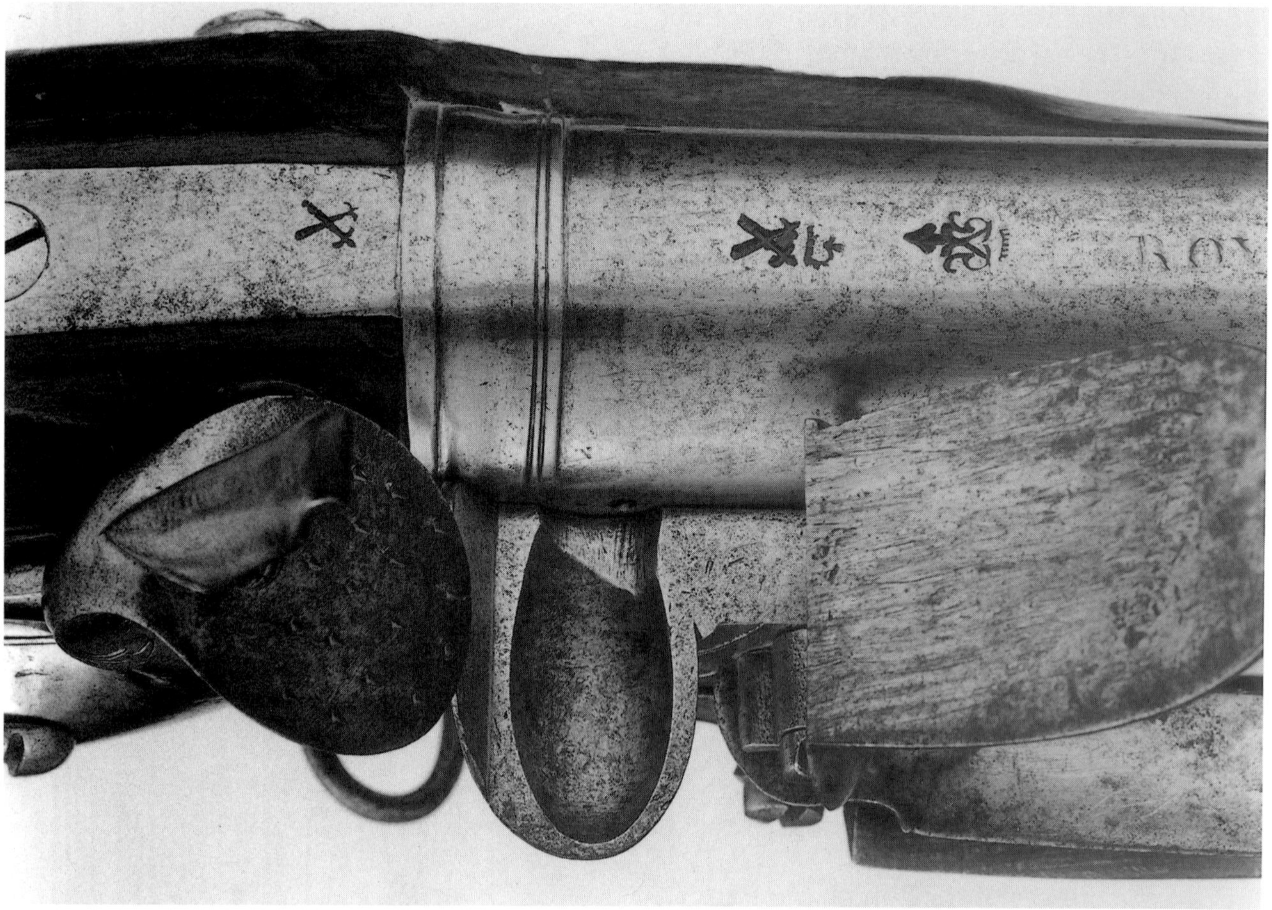

Detail of the breech of Captain Bernard musket No. 4 showing the Ordnance proof marks and the inside of the pan. This view draws attention to an interesting alteration to the pan, which has been ground back to expose the touch hole. This suggests that the lock might be a period replacement, or was originally made as such (a possibility since barrels were vented before assembly). The top jaw and screw are missing. The Colonial Williamsburg Foundation.

one of which is "Royal Welsh" marked, and has a *Tower/1741* engraved lockplate. This arm is easily distinguished from the Pattern 1730 by the substitution of an improved double bridled lock for a single-bridled one. This example, for Capt. Hickman's "H" Company, is discussed along with the true Pattern 1730 muskets, since it was certainly part of the 1741 issue. It is also set aside as unique in the group because it has the early 1740s-styled Ordnance markings on the barrel, and the more bulbous, rounded buttstock comb of that decade. This example also serves to illustrate two points. Firstly, that the arms authorized for issue in July 1741 were delivered to the Regiment late enough for a musket incorporating the latest improvements to have gone into, and be issued from, Ordnance stores. It also shows that during the pressing times of war, the Ordnance system could produce a musket with a lock bearing the same date as the weapon was assembled and authorized for issue.

In conclusion, the evidence points to the 1739 and 1741 Royal Warrants as being responsible for the issue of the eight Royal Welsh Fuzilier marked Pattern 1730s (including the lone Pattern 1730/40) known to have come from Flixton Hall. These arms went with the Regiment to the Continent, and were marked, or were re-marked, by different engravers on a company level, possibly by order of the newly appointed Colonel, General Huske.[65] Since the Regiment surrendered all of their arms from these issues in 1745, it would seem that these weapons had been in store for one reason or another when the Regiment was taken at Ghent. Perhaps they were reserves, or had been out for repair. Regardless, they saw service (some possibly at Dettingen and Fontenoy!), escaped surrender at Ghent, and found their way into the hands of William Adair after the Regiment returned to England.

Furthermore, perhaps these eight survivors were deemed surplus and/or obsolete after the Ordnance re-armed the Regiment upon their return to England, and were simply acquired by William Adair instead of being turned in. Had the 1745-issued arms been Pattern 1730/40s or Pattern 1742s, then this may have well been the case. The Regiment was to be sent back into action on the Continent shortly thereafter, and it was the practice of the Ordnance to equip such units with the latest model of weapon.

The initials of Captain William Hickman branded into the stock of a Pattern 1730/40 musket issued to his Company of the Royal Welsh Fuziliers. This transitional musket is identical to the Pattern 1730s with the exception that it was manufactured with a 1741/TOWER lock with a bridle on the pan (the true Pattern 1730 has a bridle-less pan). The Colonial Williamsburg Foundation.

Non-Royal Welsh Fuzilier Muskets, Patterns 1730, 1730/40 & 1742

The only Pattern 1730 in the Adair Collection not marked to the Royal Welsh Fuziliers is engraved on the barrel *COLL; DALLWAY'S REGT; No; VIII.*[66] The lock is marked *Es=Cole/1728* and the wristplate is engraved *Co;VIII* over *No;20*, without a line separating them. Colonel Robert Dalway commanded the 39th Regiment for a very short span between June 1739 and December 1740. Unfortunately, the manuscript Army Lists at the PRO (WO/64) do not list agents, so it can not be determined if William Adair was serving Col. Dalway in that capacity.

This musket is apparently somewhat of a "red herring." According to Ordnance records, the 39th Regiment was issued with iron rammered muskets in 1735, and was not re-armed until a warrant dated 3 December 1745 was drafted. Given that this musket was clearly never equipped with an iron ramrod, it cannot be associated with the 1735 issue. All than can be stated is that somehow, this musket got into the hands of Col. Dalway's Regiment, and was marked to the unit between June 1739 and December 1740, on a non-Ordnance level.

Another variant of the Pattern 1730/40, not regimentally marked, was in the Adair collection, and differs significantly from the "Capt. Hickman" musket formerly discussed. While it has some of the traits of the Pattern 1742 type, it retains the earlier style trigger guard and single-bridle lock. Its lockplate is marked *FARMER/1742*, and the only unit marking is the "rack" number D/48 on the wristplate.

The only other non-regimentally marked musket from the Flixton Hall group is a single Pattern 1742. It differs from the Pattern 1730s in that it has simplified carving around the lock, sideplate and barrel tang areas, a double bridled lock marked

TOWER/1742, and an improved trigger guard that has an extra inward curl behind the trigger. As with the Pattern 1730s, this Pattern 1742 is set up for a wooden ramrod, and its wristplate is very crudely engraved marked *CR/No.3*, a marking certainly not executed by an Ordnance engraver at the Tower. Since this weapon was never fitted with a noseband of any sort, it can be assumed that it was not a Royal Welsh Fusilier musket, since their muskets all had this improvement.

Regimental markings on the barrel of the only musket in the Adair Collection marked to a unit other than the Royal Welsh Fuziliers. Robert Dalway was Colonel of the 39th Regiment from June 1739 until December 1740. William Ahearn Collection.

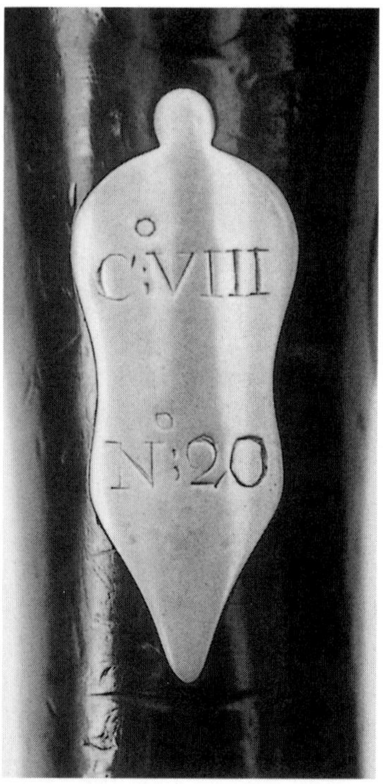

Company and weapon numbers engraved on the wristplate of the Col. Dalway musket. In style, the engraving is almost identical to that on the wristplates of the Group II Royal Welsh Pattern 1730 muskets. William Ahearn Collection.

The name of the lock maker on the Col. Dalway musket. William Ahearn Collection.

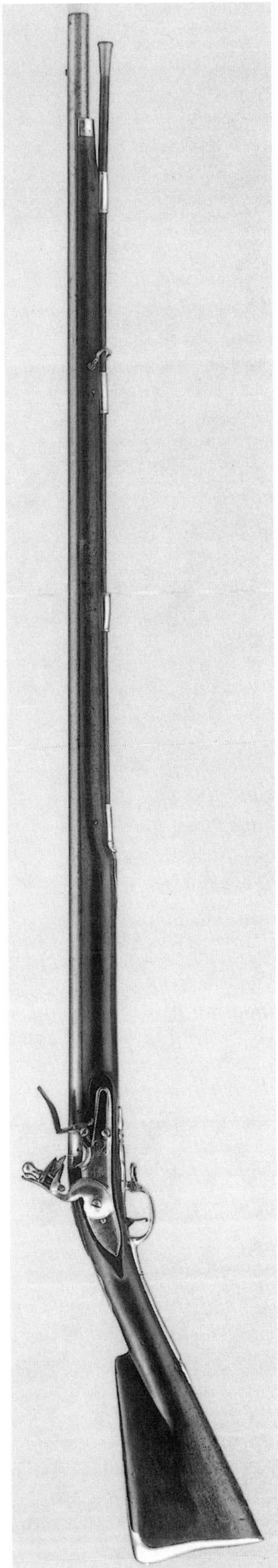

(Left) Overall view of the Pattern 1730/40 transitional musket marked "D/48" on the wristplate. The Colonial Williamsburg Foundation.

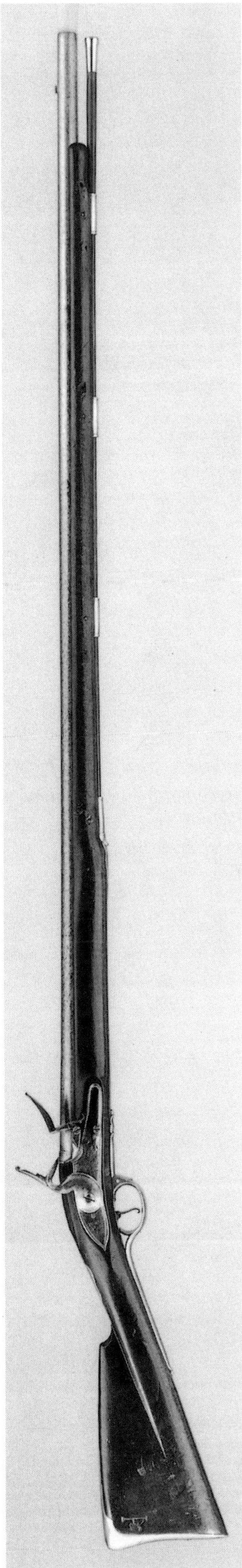

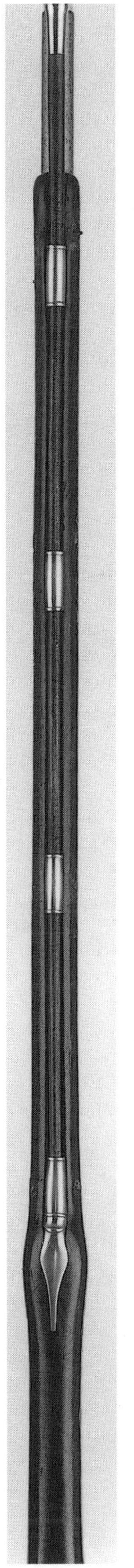

(Right) Overall view of the single Pattern 1742 musket from Flixton Hall. Its lock is marked 1742/TOWER, and it has the pan bridle. In addition to being one of two muskets from the Adair collection without regimental markings, it is the only specimen which has not been fitted with a noseband. Its wristplate is engraved "CR/ No. 3," and an "I" is carved into the left side of the buttstock. The top jaw, its screw, and the sling swivels are missing. The Colonial Williamsburg Foundation.

(Far right) Forend of the Pattern 1742 musket showing the original wood ramrod, lack of nosecap, and swell about the tail pipe. On unaltered muskets made before the mid-1740s, this swell is more prominent (see page 78, left). The Colonial Williamsburg Foundation.

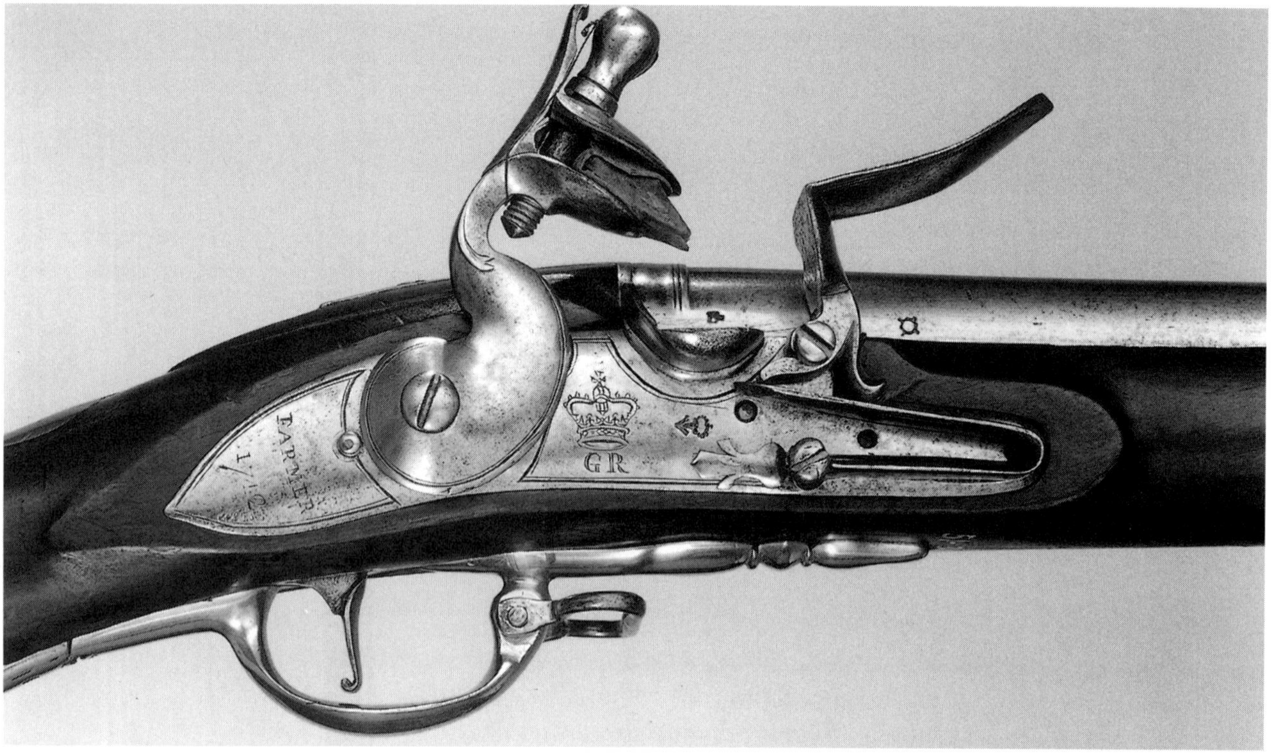

1742/FARMER lock in a Pattern 1730/40 transitional musket. This musket differs from the other Pattern 1730/40 musket in the collection in that it has a lock with a bridle-less pan (the latest dated Ordnance one observed without this improvement). It was also made without the raised carving around the lock, sideplate and barrel tang areas. Note the various markings on the breech. This is one of the two muskets from Flixton Hall that bear no regimental markings. However, the wristplate is engraved "D/48." The Colonial Williamsburg Foundation.

Muzzle of the Pattern 1742 musket, showing the original configuration of the bare forestock end, and the tip of the wooden ramrod. The slight flaring towards the end of the ramrod tip was caused by use. The Colonial Williamsburg Foundation.

The Arms Issues of 1756 and 1757

The majority of the collection of Royal Welsh Fusilier marked muskets preserved at Flixton Hall were issued to the Regiment over a period of almost exactly one year, early in the Seven Years War. By this time the Regiment had been given a seniority number, based on the date it was raised. Therefore, from 1751 on, the Royal Welsh Fuziliers were also officially known as the 23rd Regiment of Foot.

There are at least eighteen muskets from these issues known, broken down into two distinct groups. Eight are Pattern 1748s issued to the Second Battalion Royal Welsh Fuziliers and ten are Pattern 1756s issued to the First Battalion.

Just as remarkable as the survival of this group of muskets is the fact that they have been preserved with the numbered ramrods and bayonets originally issued with the group. In most cases, the bayonet survives with its original musket of issue. Considerably less frequently the ramrod survives in its original musket. Since each musket was originally issued with its numbered ramrod and bayonet, it can be assumed that at some point they simply got jumbled up. This happened long ago, since some ramrods, numbered to surviving muskets, exhibit a pitting pattern indicative of having spent long periods of time in muskets other than the ones they are numbered to.[67]

A survey of the surviving muskets, bayonets and ramrods of the 1756-7 issues shows that there is a high percentage of orphaned components. In total, there are 29 or 30 different "rack" numbers accounted for. This suggests that the original group amassed by William Adair in the late 1750s contained at least as many complete stands of these arms, and probably more (in addition to the earlier patterns). This is very compelling evidence that a significant percentage of the original collection was destroyed in the fire of December 1846.

When the bulk of the collection was sold to Colonial Williamsburg in late 1950, the group was still jumbled, and mismatched components or strays will be encountered elsewhere. For instance, the ramrod for the Pattern 1756 musket numbered 10/71, now at the Royal Armouries, is in another musket from the group now at Colonial Williamsburg. Another ramrod, numbered 3/95, was sold with the iron mounted *1724/Dublin Castle* musket in the Neal Collection sale in 1995. Other components have since been reunited with their respective muskets, and it is hoped that in the future, more such reunions will be possible.

During William Adair's time, it was common practice in the British Army for company rolls to carry non-effectives known as "contingent men." Since these men did not exist, their pay was drawn by the company commander and used towards the upkeep of the companies' weapons, amongst other contin-

The Governor's Palace at Colonial Williamsburg, c. 1994. This magnificent building is home to twenty-four of the Royal Welsh hangers and 10 of the muskets & bayonets from Flixton Hall. The Colonial Williamsburg Foundation.

gent expenses. A typical use of these "contingent" monies was the post-1743 application of markings and nosebands to the muskets issued to the Regiment in 1739 and 1741, as previously discussed. Since each company at the time consisted of seventy privates and three corporals, their weapons would have been numbered 1 through 73, with the corporals taking the first three. Therefore, the last muskets, in this case numbered 71, 72 and 73, were extras.

Although these extra "contingent" muskets could be used as seen fit by regimental powers, the company commander was still responsible for their account. As the regimental agent, the official middle-man between the Royal Welsh Fuziliers and the Board of Ordnance, William Adair was in a perfect position to acquire some of these "extra" muskets, which was apparently just what he did. All he had to do was compensate for their "loss" monetarily, no problem for a man of his wealth.

The above reason accounts for the superb, almost new state in which these muskets have been preserved. Since they never made it into the rough hands of the 23rd Regiment's soldiers, they never saw the hard use to which army muskets were subjected. Over two centuries of age and handling, or mishandling, are the only causes of the imperfections or damage seen today.

The Entrance Hall of the Governor's Palace, c. 1980-90. The two pairs of muskets immediately flanking the gilded Royal Arms over the fireplace are Pattern 1748s and Pattern 1756s from Flixton Hall. At the extreme right are some of the Royal Welsh hangers. The Colonial Williamsburg Foundation.

Pattern 1748 Muskets for the Second Battalion Royal Welsh Fuziliers

On 25 August 1756, it was decided that the Royal Welsh were to be one of fifteen regiments to be augmented by a second battalion. Therefore, each battalion was to have a total "on paper" strength of 815, incorporating ten companies of seventy privates and three corporals. While this was happening, the Regiment was still in Gibraltar, where the French had conveyed them after the Minorca capitulation in June of 1756.

To equip the new Second Battalion, a Royal Warrant dated 22 September 1756 was drafted, issuing 730 steel-rammered muskets for their use.

From this warrant, at least eight unissued Pattern 1748 transitional muskets survived at Flixton Hall. This pattern is a very seldom-encountered one and, in fact, the ones under discussion may be the only known examples. It shares traits with both Patterns of 1742 and 1756. The lock is of the curved, or "banana" profile, and has the simplified carving around the lock, sideplate and barrel tang areas. The identifying feature of the Pattern 1748 is the steel ramrod, for which it was originally set up. It has a typical Land Pattern tailpipe and three equal barrel-shaped pipes, which are all smaller in diameter than those used on the Pattern 1742 and earlier muskets set up for wooden rammers. The cap at the end of the forestock is of cast brass, as opposed to a sheet brass band.

The locks are all dated 1747, 1748 or 1750 and five carry the name of Farmer of Birmingham. One example has lost its original lock, and another probably had its lock replaced. This latter musket was not locatable for examination. The remaining example is engraved "Wood," an unrecorded contractor, and is one of the few surviving examples of his work.

All of these muskets have had their barrels beautifully engraved with the devices of the Regiment, being the Prince of Wales' plumes, coronet and motto over the regimental number "R=23."

Since they were the second of two battalions of ten companies each, the new companies were numbered 11 through 20, with 11 being the elite Grenadier Company. All of the muskets are weapon numbers 71, 72 or 73, for the "contingent men" of each company. Four of these still have their original numbered bayonets, and of these, two muskets, 17/73 and 20/72, still retain their original numbered ramrod.

1750/FARMER lock in Pattern 1748 musket 20/72 issued to the Second Battalion Royal Welsh Fuziliers in September 1756. This is the latest date for the "banana" shaped lockplate, and is one of the few muskets from Flixton Hall that retains its original numbered ramrod and bayonet. The Colonial Williamsburg Foundation.

Number	Lock marks	Bayonet	Ramrod
11/72	-	-	Yes
11/72	-	-	Yes
12/72	Wood/1748	Yes	-
13/72	-	Yes	-
13/73	Farmer/1750	-	-
17/73	Farmer/1747	Yes	Yes
19/71	Farmer/1747	-	-
19/72	Farmer/1747	Yes	-
19/73	-	-	Yes
1?/7?	Cookes/1729[1]	?	?
20/71	(replaced)[2]	-	-
20/72	Farmer/1750	Yes	Yes

1- This musket appeared in Darling's *Redcoat and Brown Bess* on page 17, Fig. C and page 18, Fig. C. From the photo, the faint devices of the Regiment can be seen on the barrel, and the wrist plate is engraved with a company number that starts with a "1" and a weapon number that starts with a "7." The lock in this musket is certainly a replacement, since a 20-year-old obsolete Pattern 1727 lock would not have been built into to a Pattern 1748 musket (which was made with the Pattern 1740 lock). Unfortunately, the whereabouts of this musket is unknown, so it has not been examined. Perhaps this Pattern 1748 musket, and the Pattern 1730 musket (Capt. Bernard N.4) now at The Royal Welch Fusiliers Museum, had their locks switched?

2 - The lock now in this musket replaces the missing original, which was marked FARMER/1747.

The Second Battalion of Royal Welsh Fuziliers was short lived, however. On 15 June 1758, they were ordered to separate from the First Battalion and become the 68th Regiment of Foot under the command of their Colonel, John Lambton. Since they were not issued a new set of arms then, it must be assumed that they continued to carry their 23rd Regiment marked Pattern 1748 muskets until they were worn out.

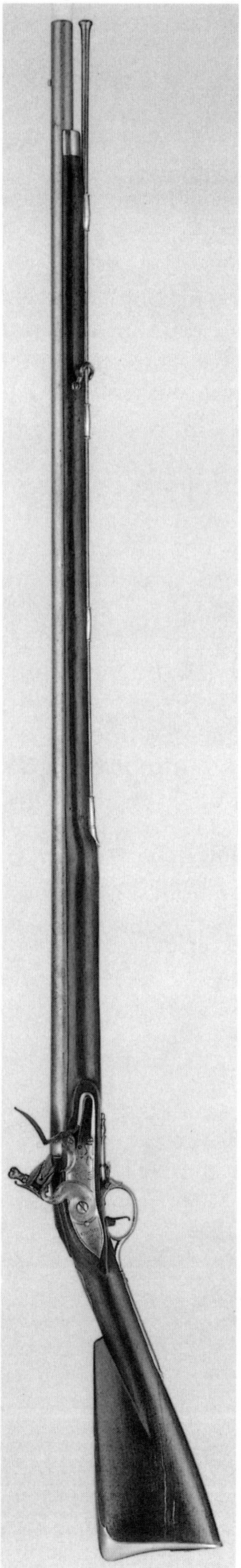

Overall view of musket 20/72. The Pattern 1748 was the only musket to be made with three equal barrel shaped ramrod pipes, a cast brass nosecap and a steel rammer. Although similar in form to the ramrod pipes on the earlier patterns, these pipes are of smaller diameter. The stock of this musket is atypical in that it has a slight "fishbelly" to the buttstock, and the areas around the lock and sideplate are not as well formed as on those usually encountered. These irregularities are probably due to variances in manufacture. The Colonial Williamsburg Foundation.

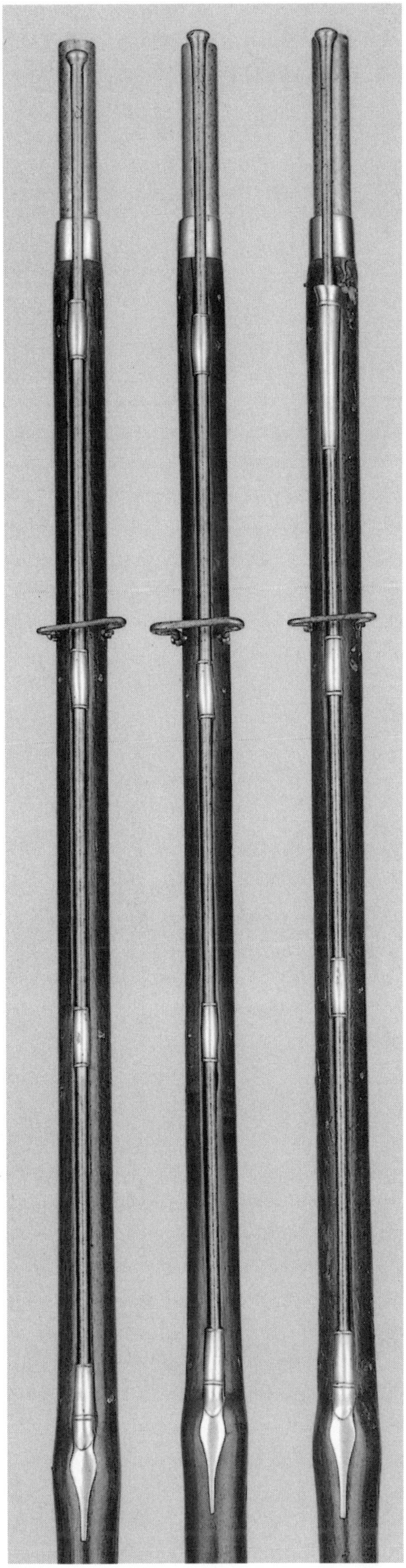

(Left) The forends of two Pattern 1748s and a Pattern 1756 musket showing the difference in the upper rammer pipes. The Colonial Williamsburg Foundation.

(Right) Overall view of Pattern 1748 musket number 17/73, with a 1747/ FARMER lock and a more typically formed stock than that on musket 20/72. The Colonial Williamsburg Foundation.

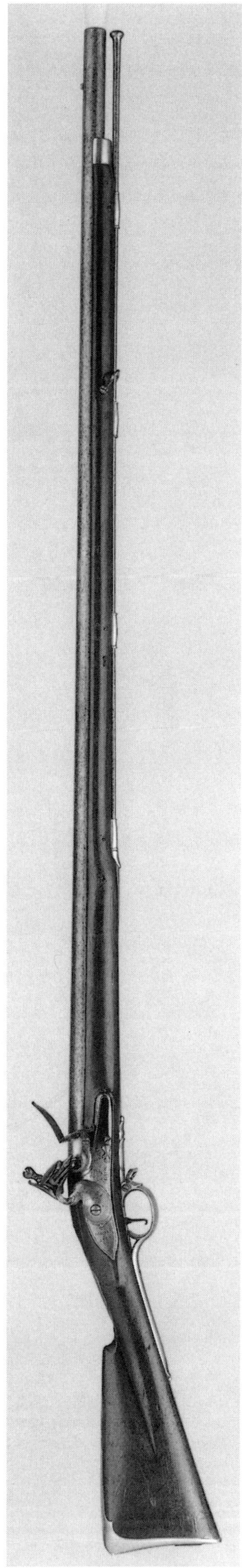

Breech detail of Pattern 1748 number 17/73, showing the engraved devices of the Royal Welsh Fuziliers. A study of the collection has shown considerable variation in the style of these engraved devices, suggesting the work of numerous engravers employed at the Tower. In addition to the usual Ordnance view and proof marks, this barrel carries the mark of Edward Jordan of Birmingham, who made the barrel, and an Ordnance inspector's mark. The Colonial Williamsburg Foundation.

Pattern 1756 Muskets for the First Battalion Royal Welsh Fuziliers

In the spring of 1756, the Royal Welsh Fuziliers formed part of the besieged garrison of Fort St. Philip on the island of Minorca. When the French successfully took the Fort, the Regiment found themselves prisoners. After being released with honors for their vigorous defense of the Fort, the Regiment returned to England via Gibralter, landing at Southampton late in November that year.

Since the Regiment arrived home without their firearms, they were rearmed with "Firelocks of different kinds," and according to the 16 February 1757 inspection return, there were "Some iron and some wooden ramrods."[68] Consequently, they were unable to go through the manual exercise or arms firings due to the poor state of these low grade weapons.

Adjutant George Brookes, Royal Welsh Fuziliers, c. 1755. Depicted in watercolor atop a donkey in front of Fort St. Philip in Minorca, Brookes had a very long career with the Regiment. He entered the ranks in 1719 at the age of 18 and was appointed Quartermaster on 28 November 1744. He was then commissioned 2nd. Lt. & Quartermaster on 29 May 1745, 2nd. Lt. and Adjutant in 1746 and Lieutenant & Adjutant on 25 June 1747. By this time, the cut of uniforms of the Officers had changed substantially from those worn in the early 1740s (see pages 17 and 75). The blue-faced red coat with gilt buttons now has wide lapels and substantially more gold lace around the edges, and is not as voluminous. Additionally, Brookes' waistcoat is now red. Adjutant Brookes died in Minorca on 25 December 1755. Courtesy of the Trustees of the Royal Welch Fusiliers Museum.

Obviously, the issue of second and third-rate weapons to the First Battalion was a stopgap measure until proper muskets could be issued. The inspection of the Regiment on 16 February 1757 at Leicester must have been quite a spectacle for Major General the Earl of Ancram. The smartly-equipped shiny new men of the partially-formed Second Battalion must have sharply contrasted with the worn and weary, ill-armed veterans of the First Battalion. From the return of this inspection, we know that far more than just the muskets were deficient: all sorts of arms and accouterments were missing, having been lost at Minorca.

Being one of the best regiments in the army, the First Battalion was soon called upon to take the field again, this time on the Continent. All of the old arms were called in by the Board of Ordnance, and a Royal Warrant for a complete set of new muskets with steel rammers and bayonets was issued on 5 March 1757. Regardless of the date of the warrant, it appears to have taken some four months to get these arms to the First Battalion, as the 24 September 1757 inspection report states that the arms were "good, having been received in July 1757."

Early in 1757 the "on paper" strength of the First Battalion was ten companies of seventy-three men (70 privates and 3 corporals), so exactly 730 Pattern 1756 muskets were issued under the 5 March warrant. It is interesting to note that the First Battalion, being sent on active service, was given the most current model of weapon. Meanwhile, the Second Battalion, not yet fully formed, and to remain in England, was issued a set of new muskets, but of a slightly obsolete pattern a few months earlier.[69]

The Pattern 1756 has two obvious improvements over the Pattern 1748. Most noticeably is the change in the shape of the lockplate, which went from having a so-called "banana" shape to having an almost

1756/EDGE lock in Pattern 1756 musket number 4/73, issued to the First Battalion Royal Welsh Fuziliers in March 1757. The locks of this pattern have a straighter lower edge than those on previous patterns. This musket retains its original numbered bayonet and ramrod. The Colonial Williamsburg Foundation.

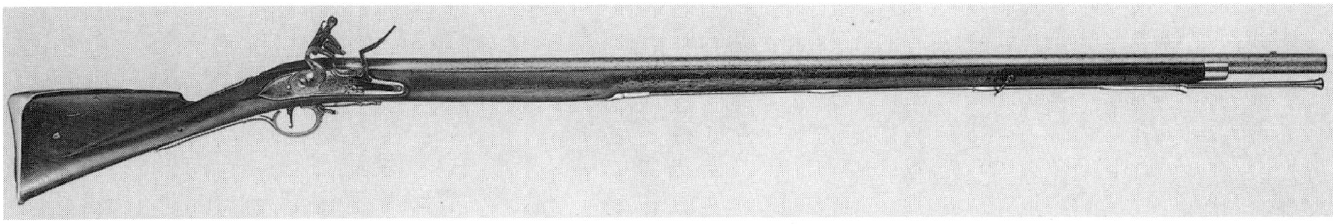

Overall view of Pattern 1756 musket number 4/73. The Colonial Williamsburg Foundation.

straight bottom edge. Additionally, a longer trumpet-top upper ramrod pipe was substituted for the short barrel-shaped one found only on the Pattern 1748.

At least ten of these Pattern 1756s survive, and are all engraved with the same devices of the Regiment found engraved on the barrels of the Pattern 1748s. The lockplates are dated 1756, with the exception of one specimen, and the makers names engraved on the tail are Edge, Grice, Jordan and Perry.

They are numbered on the wristplate exactly like those of the Second Battalion, except that the company numbers are 1 through 10. The weapon numbers are also 71, 72 or 73, for the "contingent men" of their respective companies. Surviving muskets, bayonets and ramrods from this issue are:

Number	Lock maker	Bayonet	Ramrod
4/72	Jordan	Yes	-
4/73	Edge	Yes	Yes
5/71	Perry	Yes	-
5/72	Edge	Yes	-
5/73	Grice	-	-
7/71	Edge	Yes	Yes
7/72	-	-	-
8/73	Edge	Yes	-
9/71	-	Yes	Yes
9/72	-	Yes	Yes
10/71	Grice	-	Yes
10/72	Edge	Yes	-

As often happens during wartime, in late September 1757, the Regiment was once again ordered to expand, and reform each battalion into nine companies of 100 men (plus four corporals), raising the "on paper" strength to two battalions of 936. To arm these new men, a Royal Warrant of 23 September 1757 ordered an additional 412 muskets and bayonets to be supplied for these new men (206 per battalion).

Of this September 1757 augmentation issue, only two muskets, three bayonets and four ramrods have been identified, all numbered for companies in the First Battalion. Unlike the Pattern 1756s issued to that battalion six months earlier, musket 5/94 has a 1757 dated lock. Although it is unusual for a musket's lock to bear the date of both the entire gun's assembly and issue, such is also the case of the "Capt. Hickman" No. 16 Pattern 1730/40 musket previously discussed. This phenomenon is certainly ascribable to the demands of wartime.

Although musket 5/94 retains its original ramrod, it has unfortunately been separated from its bayonet (current whereabouts are unknown). Bayonet 5/94 was once in the collection of New Jersey collector Jac Weller, and was sold by Butterfield & Butterfield (of San Francisco, CA) as part of lot 2360 in their 16 May 1995 auction of Weller's collection.

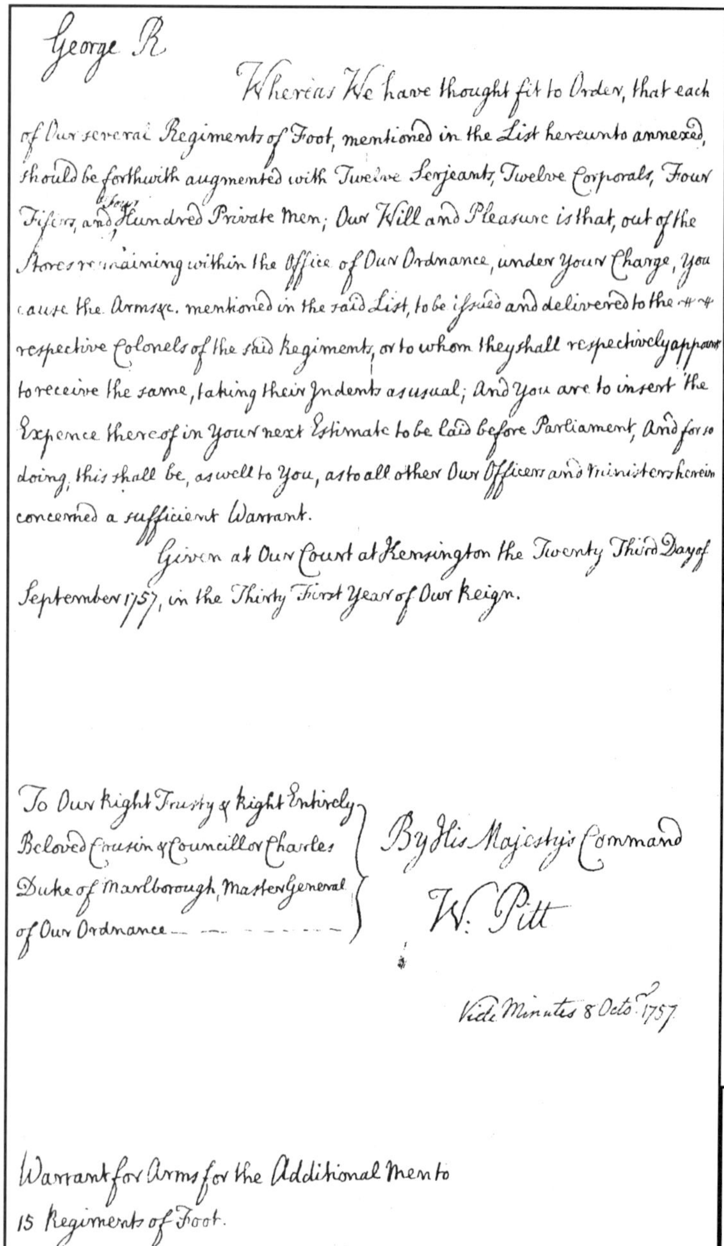

George R

Whereas We have thought fit to Order, that each of Our several Regiments of Foot, mentioned in the List hereunto annexed, should be forthwith augmented with Twelve Serjeants, Twelve Corporals, Four Fifers, and Four Hundred Private men; Our Will and Pleasure is that, out of the Stores remaining within the Office of Our Ordnance, under Your Charge, you cause the Arms &c. mentioned in the said List, to be issued and delivered to the respective Colonels of the said Regiments, or to whom they shall respectively appoint to receive the same, taking their Indents as usual; And You are to insert the Expence thereof in Your next Estimate to be laid before Parliament, And for so doing, this shall be, as well to You, as to all other Our Officers and Ministers herein concerned a sufficient Warrant.

Given at Our Court at Kensington the Twenty Third Day of September 1757, in the Thirty First Year of Our Reign.

To Our Right Trusty & Right Entirely Beloved Cousin & Councillor Charles Duke of Marlborough, Master General of Our Ordnance

By His Majesty's Command
W. Pitt

Vide Minutes 8 Octo. 1757

Warrant for Arms for the Additional Men to 15 Regiments of Foot.

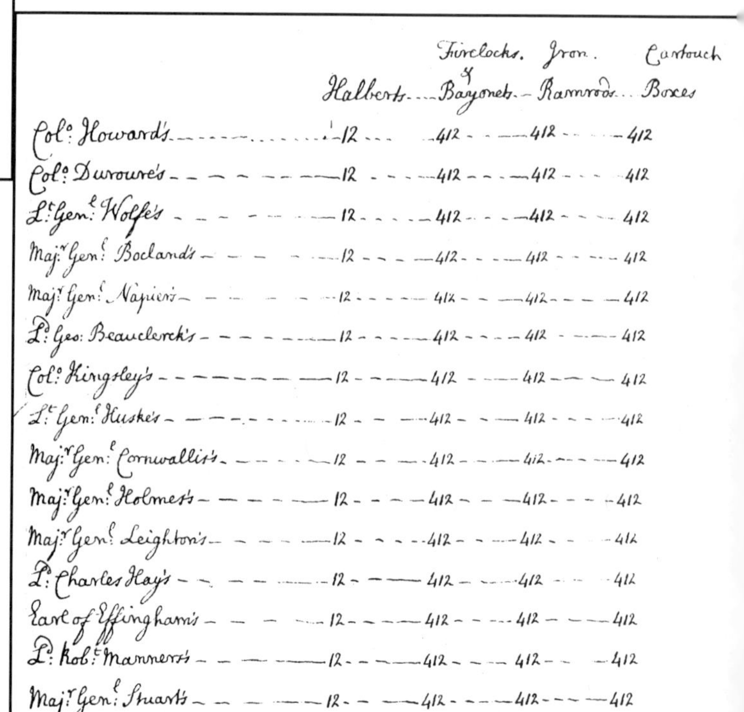

	Halberts	Firelocks & Bayonets	Iron Ramrods	Cartouch Boxes
Col.o Howard's	12	412	412	412
Col.o Duroure's	12	412	412	412
Lt. Gen.l Wolfe's	12	412	412	412
Maj.r Gen.l Bocland's	12	412	412	412
Maj.r Gen.l Napier's	12	412	412	412
Ld. Geo: Beauclerck's	12	412	412	412
Col.o Kingsley's	12	412	412	412
Lt. Gen.l Huske's	12	412	412	412
Maj.r Gen.l Cornwallis's	12	412	412	412
Maj.r Gen.l Holmes's	12	412	412	412
Maj.r Gen.l Leighton's	12	412	412	412
Ld. Charles Hay's	12	412	412	412
Earl of Effingham's	12	412	412	412
Ld. Rob.t Manners's	12	412	412	412
Maj.r Gen.l Stuart's	12	412	412	412

The original Royal Warrant issuing arms and "cartouch" boxes for the augmentation of the Royal Welsh Fuziliers, in addition to 14 other regiments, on 23 September 1757. Note that the directions specify that these items are "to be issued and delivered to the respective Colonels of the said regiments, or to whom they shall respectively appoint to receive the same," in reference to what would normally be the agent, or a regimental officer. Since the Colonels were responsible for receiving these supplies, it is their names, which appear on the list, as opposed to their regiment's title or seniority number. Under this warrant, those units receiving arms were the 3rd, 4th, 8th, 11th, 12th, 19th, 20th, 23rd, 24th, 31st, 32nd, 33rd, 34th, 36th, and 37th Regiments of foot (in order from top to bottom). Courtesy of the Public Records Office, London, War Office papers (WO 55/356, pp. 71&72).

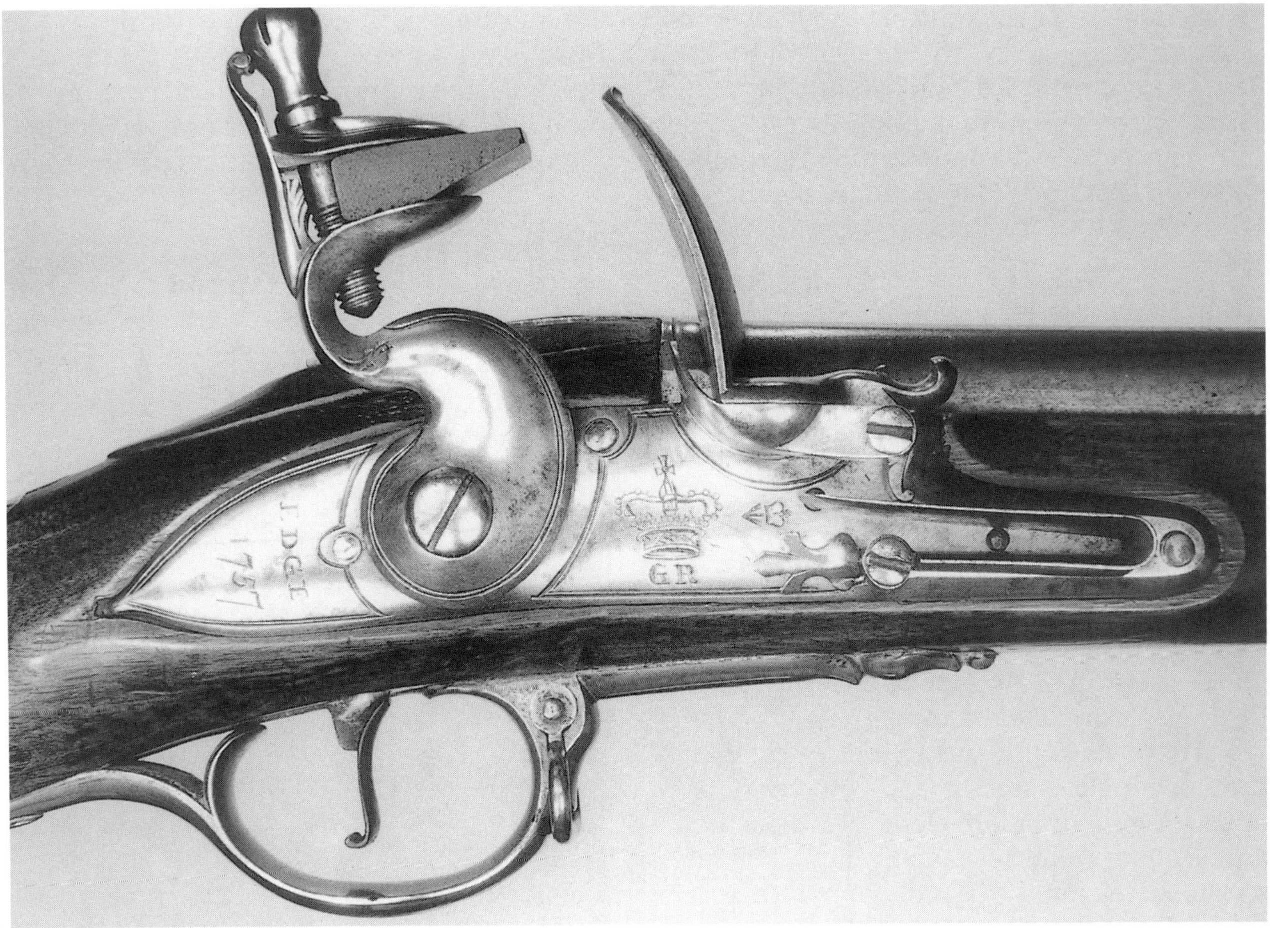

1757/EDGE lock in Pattern 1756 musket number 5/94, issued to the First Battalion of Royal Welsh Fuziliers for their augmentation of 23 September 1757. Private collection.

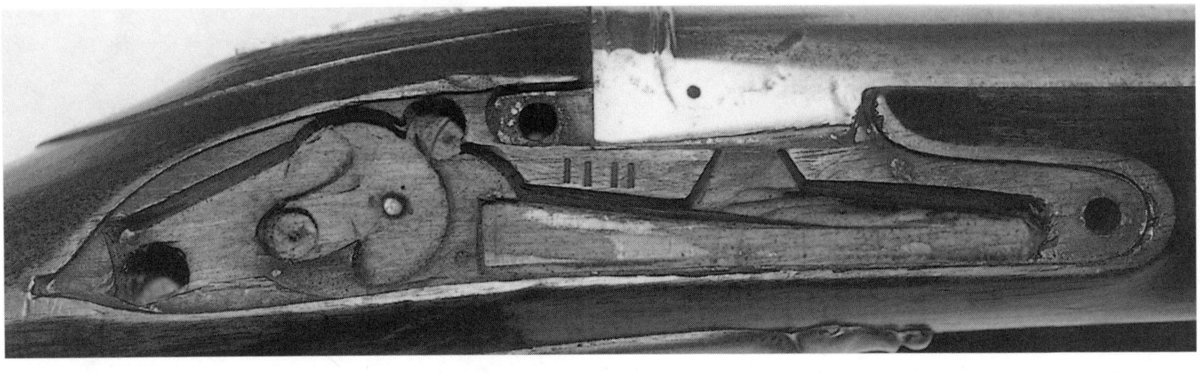

The lock mortise of musket 5/94. Private collection.

Since the muskets ordered for this augmentation were to be issued to the newly-added men of each company, the numbering system picked up where it left off. Therefore, weapons numbers ranged between 74 and 100. Since the below listed muskets, bayonets and ramrods are numbered between 94 and 96, it may well be that these men were also "contingent men." The weapons numbers for surviving components of the 23 September 1757 issue are:

Number	Lock marks	Bayonet	Ramrod
3/95	-	Yes	Yes
5/94	Edge/1757	Yes	Yes
6/96	Galton/1756	Yes	-
7/96	-	-	Yes
8/96	-	-	Yes

It can be said for certain that these two issues of muskets received by the First Battalion in 1757 were those used at the Battle on Minden on 1 August 1759, one of the Regiment's greatest victories.

There was only one further Royal Warrant for muskets and bayonets to be issued to the Regiment before General Huske died and William Adair lost the agency. For unspecified reasons, 306 muskets with steel rammers (probably Pattern 1756s) were ordered on 23 November 1759 and sent to the Regiment in Germany. Since this was only a few months after the Battle of Minden, they may have been replacements for arms lost and damaged during that campaign. No weapons from this issue have been identified in the group from Flixton Hall, and are not likely to have been part of it, since the lot was shipped straight to the Continent.

Detail of the barrel engraving of musket 5/94. Private collection.

Breech detail of musket 5/94 showing the Ordnance view and proof marks in addition to the mark of the firm of Farmer & Galton of Birmingham, who made the barrel. Private collection.

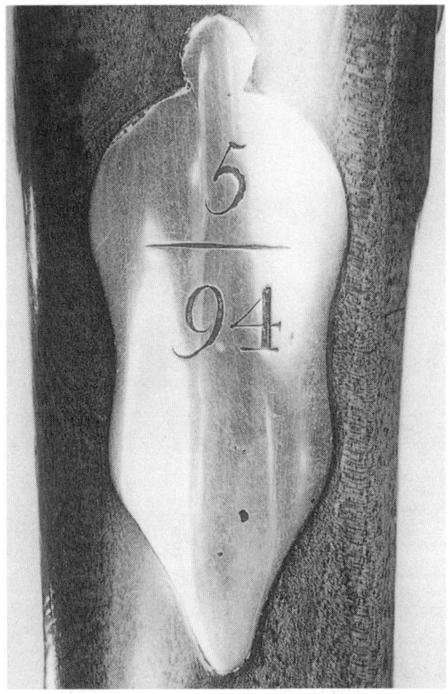

Company and weapon number engraved on the wristplate of musket 5/94. Private collection.

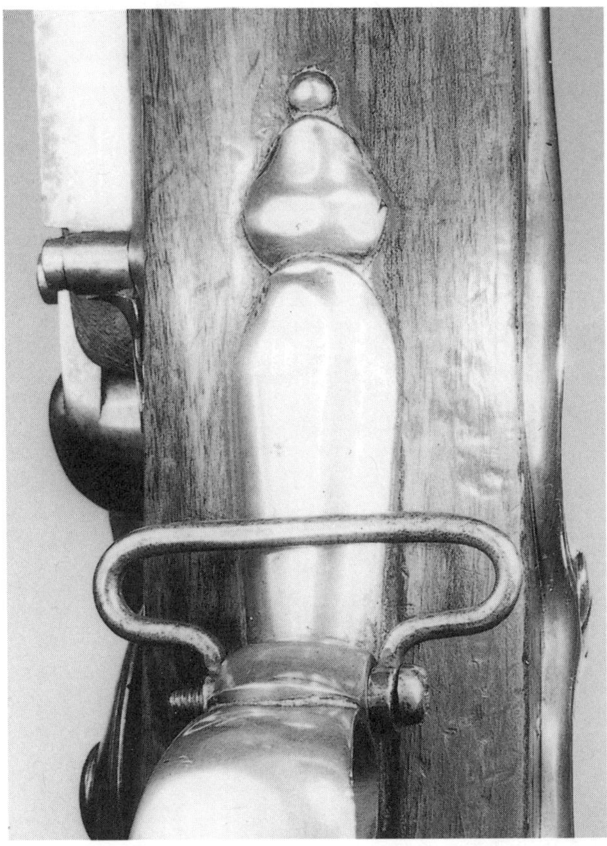

Trigger guard finial and lower sling swivel of musket 5/94. Private collection.

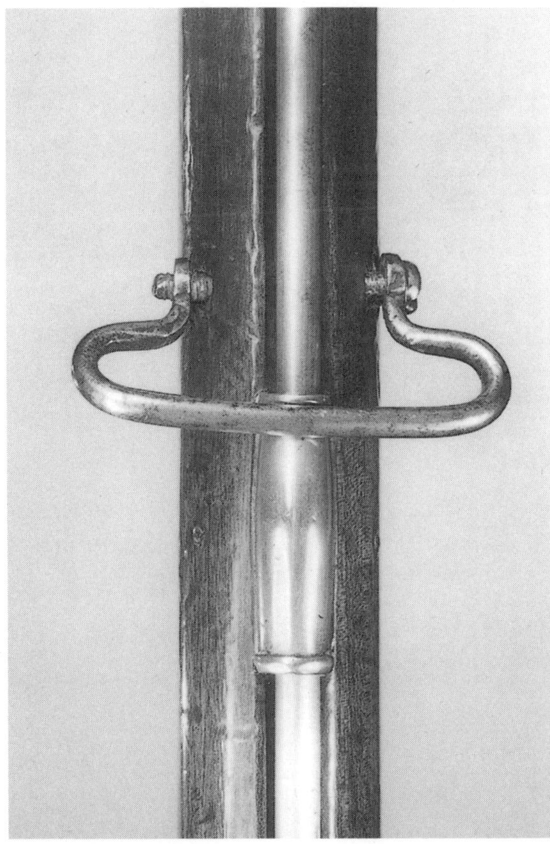

Upper sling swivel and second rammer pipe of musket 5/94. Private collection.

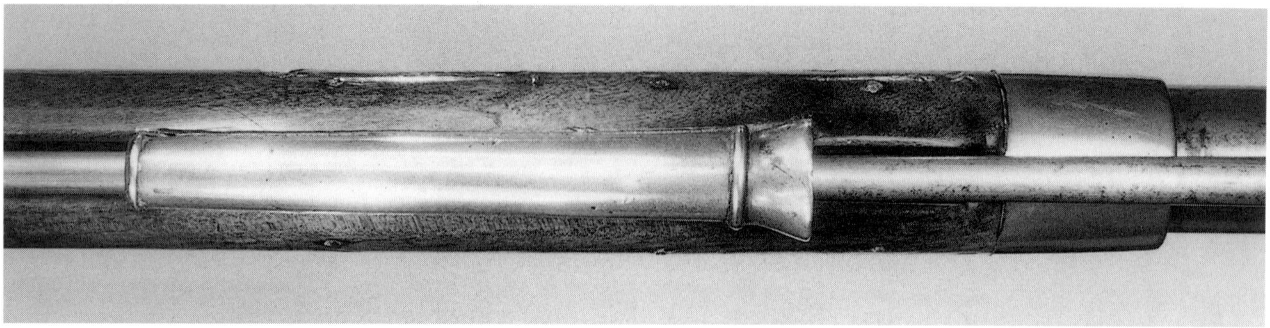

The improved longer, trumpeted upper rammer pipe of musket 5/94, as introduced with the Pattern 1756 musket. Private collection.

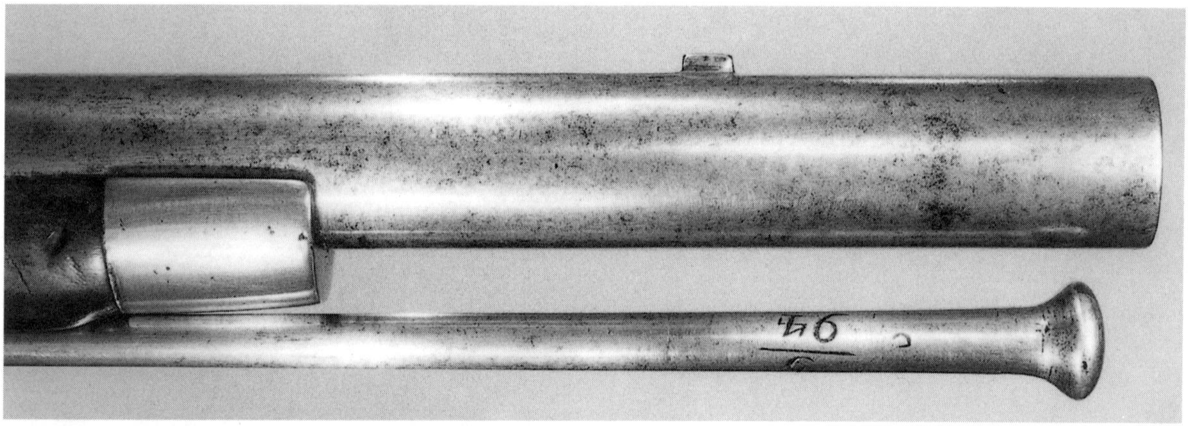

Muzzle area of musket 5/94, showing the cast brass nosecap and its original numbered rammer. Private collection.

The Bayonets

When Neal sold the bulk of the collection to Colonial Williamsburg in 1950, the muskets came with what he termed "their own bayonets." In the cases of the Patterns 1748 and 1756 muskets, this was true for the most part. In the other cases, bayonets were included with each musket, so each "stand" was complete. Unfortunately, in the cases of the Pattern 1742 and earlier muskets, the bayonets that went to Colonial Williamsburg with them are probably from the collection at Flixton Hall, but are not specifically numbered to the guns.

The Pattern 1727 bayonet is the first standardized bayonet produced in large numbers under the Ordnance System. It is characterized by a long, forward sweeping shank attached to the extreme front of the socket and is marked with a filed "shield" shaped embellishment. It is further identified by the Board of Ordnance's ownership mark of a "crowned broad arrow" stamped on either the blade or shank.[70]

The bayonets Neal supplied with two of the Pattern 1730 muskets at Colonial Williamsburg are Pattern 1727s. One example is stamped on the blade with a sunburst-like mark, and the other is marked with a "W." Both examples are marked on the shank with a "crowned broad arrow" of the type found on Ordnance lockplates dated between 1730 and 1738, thereby dating these bayonets. Neither example has a weapon or company number.

The case is very different for the Pattern 1748 and 1756 muskets issued to the 23rd Regiment in 1756 and 1757. Of the eighteen known muskets from these issues, eleven, all at Colonial Williamsburg, retain their original numbered bayonets. Until 1999, two of the bayonets for muskets at Williamsburg remained at large, but have

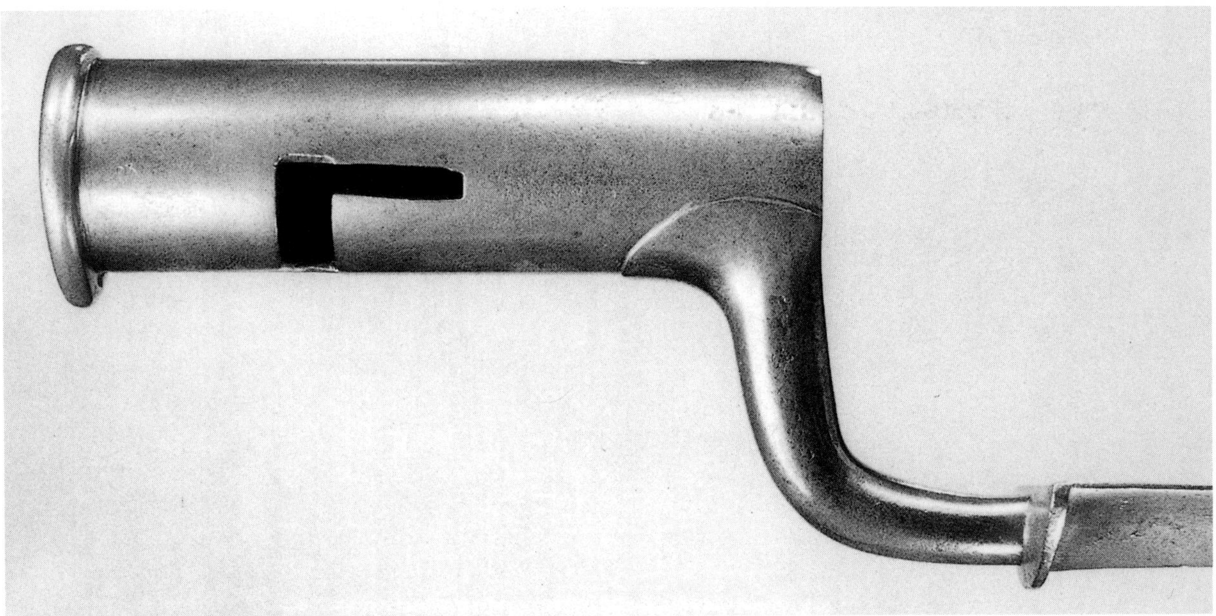

Pattern 1727 bayonet showing the characteristic long, forward sweeping shank and "shield" shaped decoration at its junction to the socket. Private collection.

Blade markings of the Pattern 1727 bayonet (previous page) showing the "crowned broad arrow" of the type found on Ordnance lockplates dated between 1727 and 1729 (see page 68, top). Similar sunburst-like markings have been observed on other Pattern 1727 bayonets (next page). Private collection.

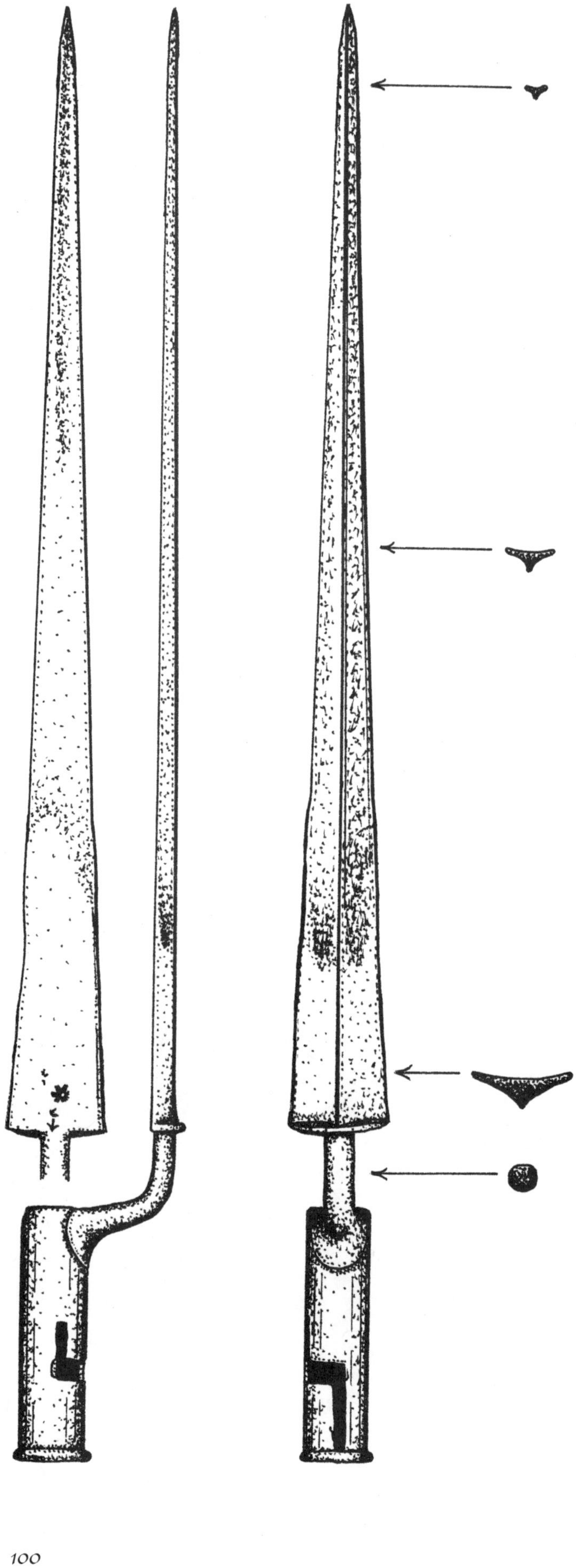

Overall view of Pattern 1727 bayonet. Socket 4", blade 17 1/2" overall length 22 5/8". Private collection.

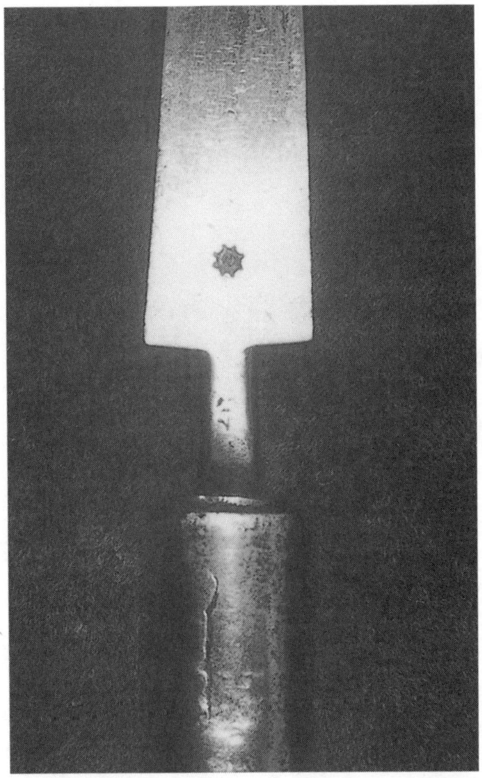

Blade and shank markings of a Pattern 1727 bayonet from Flixton Hall, with a similar sunburst-like mark to that in (photos on pages 99 and 100). The "crowned broad arrow" struck on the shank is the type seen on Ordnance lockplates dated between 1730 and 1738. The Colonial Williamsburg Foundation.

since been reunited with their firearms of issue. As of the time of this writing, a total of sixteen bayonets have been identified, and it is thought that there are others from Flixton Hall still out there, awaiting recognition. An examination of the charts in the previous chapter will show what bayonets are potentially at large.

All of the bayonets are essentially identical, being of the first generation of Land Pattern bayonet made without the "shield" shaped decoration at the socket-to-shank junction. Amongst a variety of other markings, all of the blades are marked with a "Crown 5," the mark of an Ordnance "Viewer" working in Birmingham during the c. 1740-1760 period. Since a permanent "Viewer" was initially stationed there in 1755, the "Crown 5" is most probably the inspection mark of George Markby (or one of his staff), the first man to hold this post.

These "Crown 5" bayonets have the standard Land Pattern socket and mortise setup, and a thick curved shank with a flat front face. Blades will vary in their rate of taper, but have wide shoulders in the vicinity of 1-5/16 to 1-7/16 inches across. They have a prominent base guard and have an iron and steel join about 1/3 of the way up the blade from the base.

Often the top face of the blades will have three distinct groups of markings. The closest to the tip will be a number, the second down will be either initials or a cutler's style mark, and the nearest to the shank will be the "Crown 5". Four of the bayonets from the group have been observed with a mark on the inside of the socket where the shank joins it. On two others, 5/71 and 7/71, the "Crown 5" is surmounted with a letter "P," which appears to be an integral part of the stamp, but its significance is unknown.

Most of these bayonets have survived in almost new condition, since they were never in the busy hands of soldiers. A few have suffered more pitting than others, and a few show signs of having been slightly abused at some point long ago.

It is apparent that these bayonets spent much of their time at Flixton Hall, displayed separately from the muskets, perhaps in a "bayonet fan," like those seen today at Hampton Court and Chevening House. Had they been displayed on the muskets, then the areas protected by the sockets would have received less pitting than the rest of the barrels (with the exception of the areas left exposed by the mortise of the socket). None of the muskets show anything other than an even pitting pattern at their muzzles, supporting this theory.

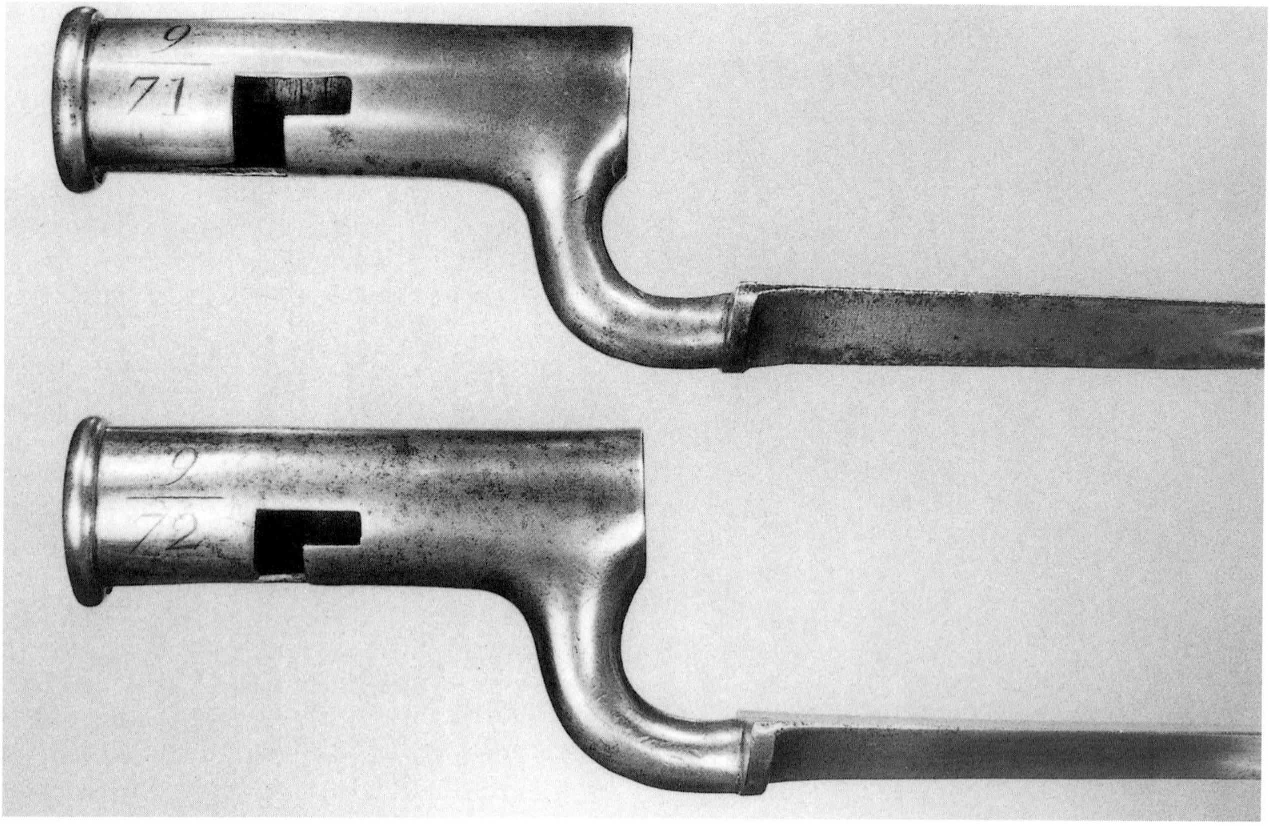

Land Pattern bayonets numbered 9/71 and 9/72, issued to the First Battalion Royal Welsh Fuziliers in March 1757. Note that these bayonets were made without the shield shaped decoration at the socket-to-shank junction. Although these bayonets are consecutively numbered in the same company, their corresponding muskets have not survived, or have not yet been located. Perhaps muskets 9/71 and 9/72 were victims of the 1846 fire at Flixton Hall. Private collection.

Blade markings of bayonet 9/72. The "Crown 5" is probably the mark of George Markby, the Ordnance inspector permanently stationed in Birmingham after 1755, although he worked there before that date. Socket 4", blade 17 x 1-7/16" overall length 21-3/4". Private collection.

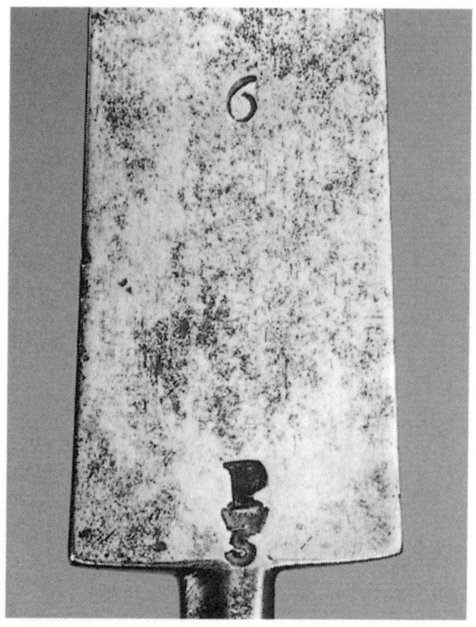

Blade markings of bayonet 5/71, also issued to the First Battalion Royal Welsh Fuziliers in March of 1757. This bayonet is one of two (the other is bayonet 7/71) from Flixton Hall where the "Crown 5" inspector's mark is surmounted by a "P." The Colonial Williamsburg Foundation.

Socket markings of bayonet 20/72 issued to the Second Battalion Royal Welsh Fuziliers in September of 1756. This bayonet corresponds to the musket shown on page 87 and 88. The Colonial Williamsburg Foundation.

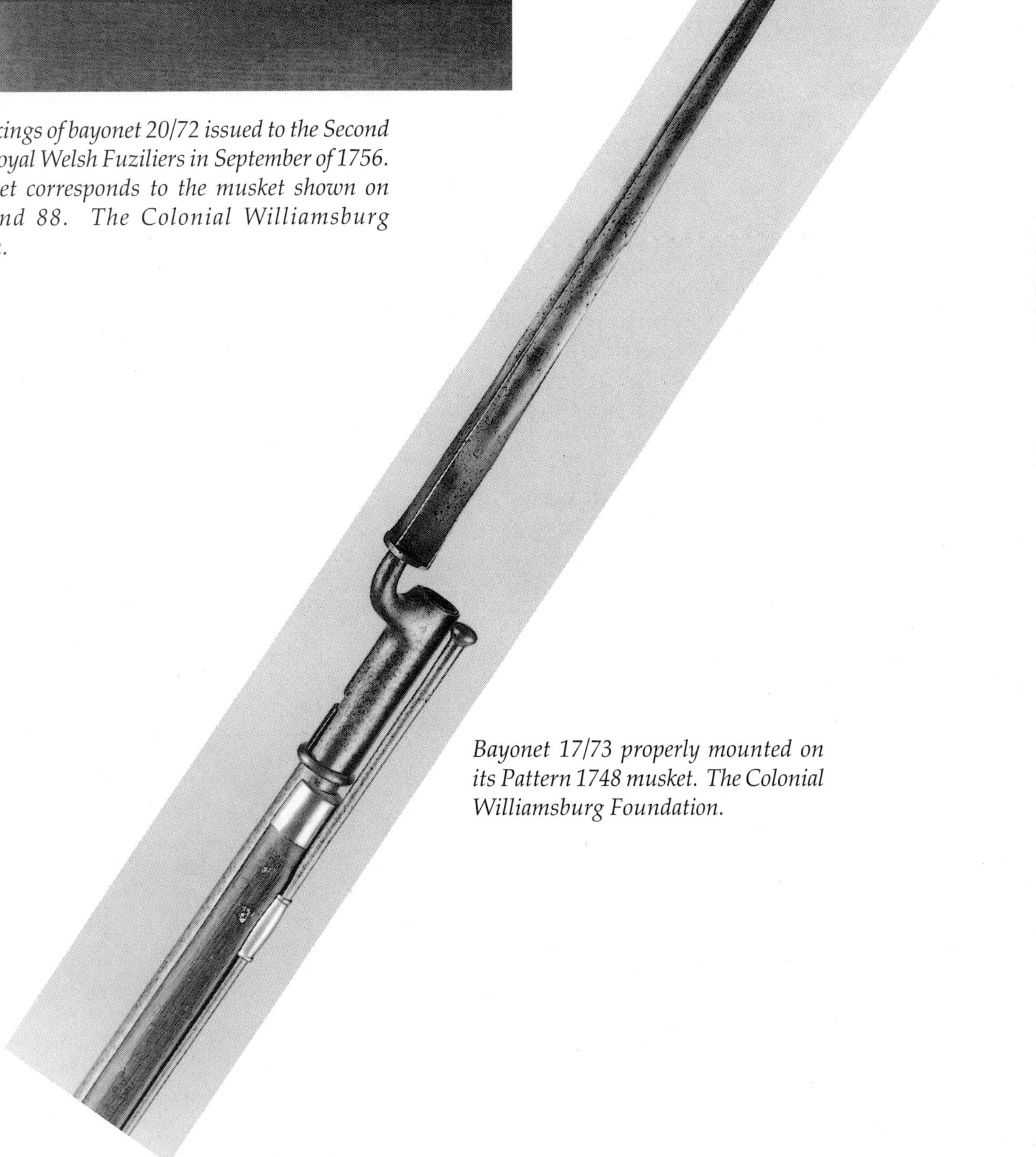

Bayonet 17/73 properly mounted on its Pattern 1748 musket. The Colonial Williamsburg Foundation.

The Swords

While weapons such as muskets and bayonets were obtained from the Board of Ordnance stores at the Tower of London, hangers were classed as accouterments and were usually purchased privately by the regiment's Colonel, through his agent.[71] Since Colonels were still allowed to equip their men as smartly or cheaply as they saw fit, the Royal Welsh Fuziliers chose the former, having their own distinct patterns of hangers from at least the early 1740s. Had William Adair's business papers survived, it would likely have been possible to learn the source, cost, and date of purchase of these specially made regimental swords.

William Keith Neal quickly realized that he had stumbled upon a group of mid-eighteenth century infantry hangers unlike any others that he had seen before. Two years after the purchase of the muskets and halberds, Neal sold the group of hangers to Williamsburg. Concerning them, on 29 December 1952 he wrote M. W. Thomas, Jr.:

> I think I did mention that these swords came from the same family stand of arms as the set of fine early long muskets which I sent you, many of them having the same regiment, Royal Welsh Fuziliers arms on the barrels. I do not think that there is another "run" of swords of this period in existence out of captivity.
>
> These swords like the muskets came from one of the greatest military families of England, and I bought them privately from the present representative, a Brigadier General, and took them down from the walls myself of his family mansion. They are more than just specimens but swords that have seen actual service.[72]

In all, twenty-six of these hangers went to Colonial Williamsburg. Of this group, one was stolen in the 1960s, and another example was exchanged and is now part of the George C. Neumann Collection at Valley Forge National Historical Park. Three others are known in North American collections, and a few more reportedly exist in others. Since Neal sold some of the muskets to associates, as well as Colonial Williamsburg, it seems likely that some of the hangers also went elsewhere. One ex-

Hilt of one of the Royal Welsh hangers in a photograph sent by W. Keith Neal to Colonial Williamsburg with his letter of 2 November 1950. Note the fine, crisp detail of the devices on the grip. The Colonial Williamsburg Foundation.

ample, on page 106, was purchased by a previous owner in England in the 1950s. Therefore, it seems possible that these other hangers without known provenance also originated from Flixton Hall.[73]

These hangers are of a form very common to the British armed forces of the period, with a hilt composed of cast brass components. They have a slightly dished, heart-shaped counterguard, strengthened from the underside with a raised edge, which then tapers into the knucklebow. A lobe-shaped quillon is included for extra wrist protection and aesthetic balance. As on other contemporary hanger variations, its pommel is spherical with an integral capstan, incised decorative rings, a narrowed supporting mantle, and an open foot that fits over the top of the grip.

The grip displays the Royal Crown over the Prince of Wales's device over the Horse of Hanover passant. It is composed of two identical castings brazed together along the sides and filled with a wood core. Both halves of the grip are made from the same mold or matrix, evidenced by design consistencies present on all known examples. The background areas are textured like an orange peel, likely intended to improve grip. Some specimens have very crisp detail, while others appear very worn, poorly cast, or both.

Since the hilts of these swords are simple and their blades are short, it can be assumed that they were for the battalion companies. Grenadier companies tended to carry either basket or semi-basket hilted hangers with blades that were longer, in the 27 to 31 inch range. Additionally, the sword worn by the Royal Welsh Grenadier in Morier's c. 1751 painting wears a sword which is completely different from those in discussion.

Two very similar specimens, certainly not from Flixton Hall, have a background textured with neat rows of stippling. Since this grip variation displays the same design characteristics as the others, it was likely produced from the same mold or matrix as the others, either after or before reworking.

A survey of twenty-eight specimens shows that the blades measure between 23 and 25-1/2 inches in length, average about 1-1/8 inches wide, and are invariably slightly curved with a short false edge and no fullers. The blades also exhibit an iron-to-steel joint a few inches from the hilt.[74] Many specimens have heavily nicked blades, three have replaced quillons and one is missing much of its knucklebow. Others show the old sharpening and re-pointing mentioned by Neal in 1953. All of this points to use and wear, in contrast to the

The Royal Welsh Fuzilier hanger and scabbard over the northern-most arch between the Flixton Hall's grand staircase and great hall (a detail of photo on page 55). This is the only image found which clearly depicts a weapon of the Regiment on display in the mansion. Courtesy of Lady Darell.

"unissued" states of the Patterns 1748 and 1756 muskets. No markings of any type have been noted on either the hilts or the blades.

Nine of the swords preserved at Colonial Williamsburg retain their original wood-lined scabbards, which are covered with tooled black leather, stitched down the center of the rear side. An elongated iron hook is set near the mouth on two specimens, and the tips are protected by sheet brass chapes decorated with finials, notches and incised lines.

Royal Welsh Fusilier regimental records of the period have made occasional references to hangers, but none have yet been found that give clues as to what they looked like. During the siege of Fort St. Philip on Minorca, on 27 June 1756, the Royal Welsh were part of a working party, apparently quartered outside of the main fortification. That night at 10:00 P.M., the French assaulted the "Inner Covert (sic) Way," but were repulsed. Unfortunately for the Regiment, their quarters were taken, and "most of the men lost their knapsacks, waistbelts, pouches, swords, etc."[75] It is unclear as to whether these items were destroyed in the action, or carried off as spoils by the French forces.

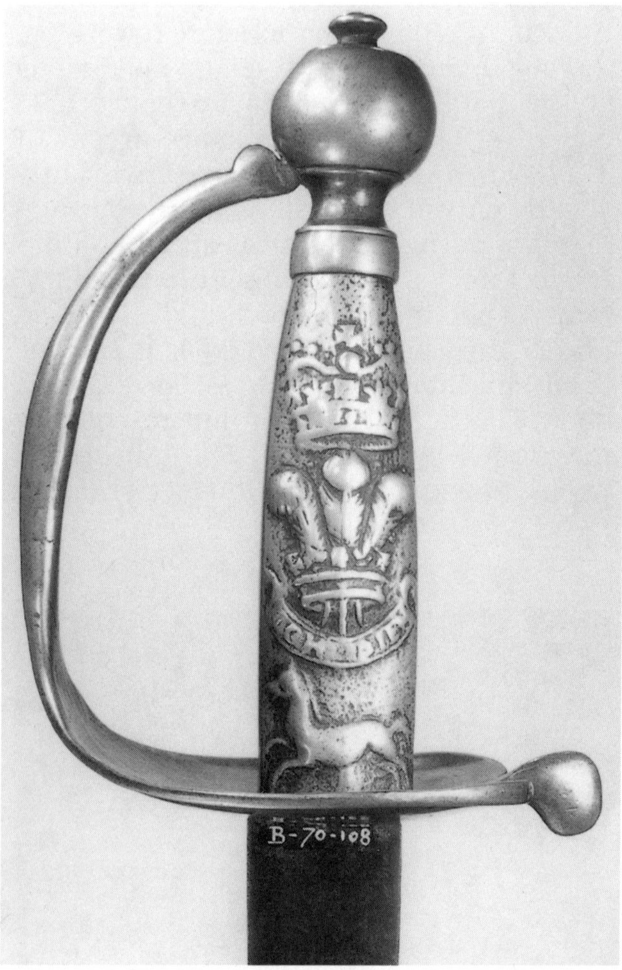

Hilt of a Royal Welsh hanger with less detail in the Regimental devices than that on page 104. This is likely due to poor casting, extensive wear, or a combination of both. Additionally, the pommel is a bit compressed and dented. Hilt 5-7/8", blade 23-1/4" overall length 29-1/8". Private collection.

This is further confirmed by the return from the inspection of the Regiment conducted on 16 February 1757 by Major General the Earl of Ancram, which states:

> "Return of Accoutrements"
>
> "Colors 2 bad. Wanting 662 swords, 206 waistbelts, 577 pouches with shoulder belts, 299 slings, 35 grenadier match cases, 7 pioneer axes, 3 saws, 10 aprons. The working men the night of the attack lost all of their pouches and almost all of their waistbelts and swords."[76]

From this document we can see that the Regiment was in need of 662 swords, to equip 700 to 730 enlisted men, N.C.O.s and musicians.[77] Since the Grenadier Company lost their "pouches" (cartridge boxes) and match cases, it can be assumed that they also lost many, if not most, of their hangers. Unfortunately, the return makes no distinction between the grenadiers' swords and those of the battalion company men.

While the First Battalion was still on Minorca, the Second Battalion was being formed at Leicester. Since the First Battalion's swords had to be replaced, and the new battalion required them, the number of required hangers was about 1,460, before the September 1757 augmentation.

Evidence exists which suggests that the needed hangers were soon procured. On 24 September 1757, Major General Lord George Sackville inspected the First Battalion at Chatham, and it was reported that all accoutrements were "well put on."

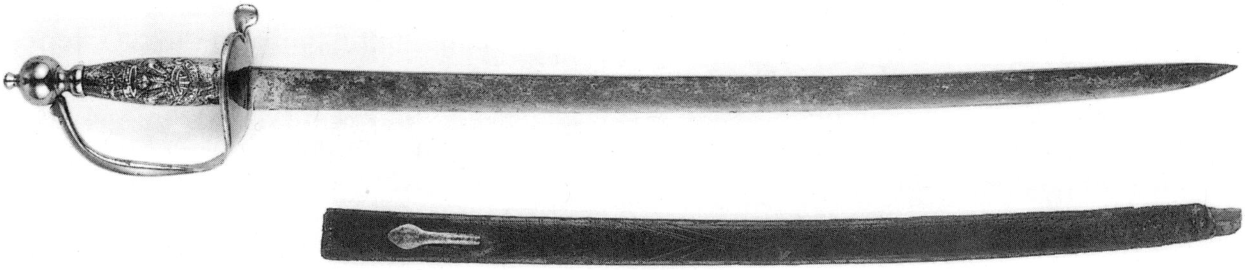

Overall view of one of the hangers with its scabbard. The Colonial Williamsburg Foundation.

View of the Entrance Hall of the Governor's Palace at Colonial Williamsburg, 1981. The twenty-four Royal Welsh hangers are visible flanking the archway at the center under the flags. The first four muskets to the left of the arch and the first six muskets to the right of the arch are all Royal Welsh Pattern 1748 and 1756 muskets from Flixton Hall. The Colonial Williamsburg Foundation.

Had their hangers been missing, surely it would have been noted, especially for a regiment soon to be sent into action on the Continent.

As with the other weapons in the collection, William Adair was in the perfect position to acquire as many of these regimental swords as he wanted. In fact, it may have been even easier to obtain these hangers, since they were the privately purchased property of the Regiment, so no accounts concerning them had to be supplied to the Board of Ordnance.

Based on their states of condition, most of these swords appear to have made it into the hands of the soldiers. Although some of the damage and wear they exhibit today may be the result of later abuse and harsh cleaning, it seems likely that when William Adair acquired them c. 1760, all had scabbards, and all were used but sound.

Royal Welsh Fuzilier private, c. 1742, from A Representation of the Cloathing of His Majesty's Household and all of the Forces upon the Establishment of Great Britain & Ireland, *otherwise known as the "1742 Cloathing Book." Note the devices on his mitre cap, which are the same as those cast into the grips of the Regiment's hangers. Courtesy of the Anne S. K. Brown Military Collection, Brown University.*

Detail from Morier's c. 1751 depiction of a Grenadier Private of the Royal Welsh Fuziliers. His mitre cap also displays these same three devices. The Royal Collection, Copyright, Her Majesty Queen Elizabeth II.

Variant Royal Welsh Fuzilier brass hilted hanger, c. 1750-60, with a different counterguard and a stippled background to the grip, certainly not from the Adair collection at Flixton Hall. Since this sword turned up in France in the late 1990s and is unlike those from Flixton, it is very tempting to associate this sword with those the Regiment lost to the French at Minorca in June 1756. Hilt 6", blade 24-13/16" overall length 30-13/16". Private collection.

Gorget of an Officer of the Royal Welsh Fuziliers, c. 1750-1770, engraved with the three devices of the Regiment as found on the hanger grips and mitre caps. This gorget measures 5-1/2 inches high and 5 inches wide. Private collection.

Comparison of the typical Royal Welsh hanger grip to the variant with the stippled background. Note the various design consistencies which indicate that both types were made from the same mold matrix. The background of the matrix was certainly reworked at some point, but it can not be said which variation preceded the other. Private collection.

Iron-hilted "Double S" infantry hanger for the Royal Welsh Fuziliers, c. 1740-1760, which surfaced during the late 1960s in England, from an unknown source. Although the grip has the same devices as the cast brass hangers, it is constructed of embossed and gilt sheet brass over a wood core. The blade measures 24-1/2 inches, and its hilt is smaller than those encountered on other non-Royal Welsh "Double S" hangers. A second example in the collection at York Castle (Collection #CA-1040) has residue of what appears to be period red lacquer on the matted areas of the grip surrounding the devices of the Regiment. The George C. Neumann Collection, Valley Forge National Historic Park.

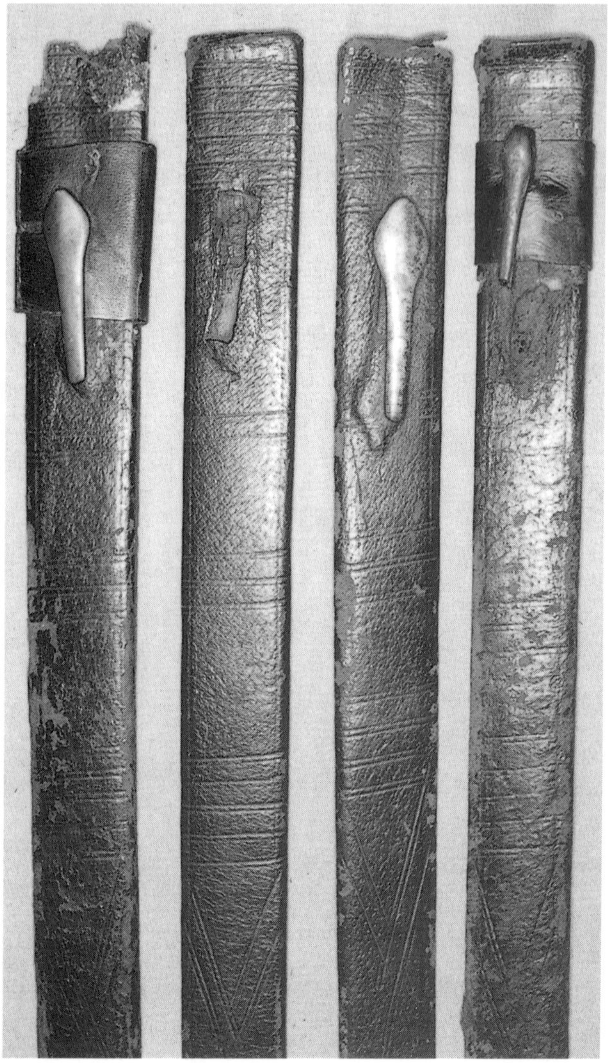

The upper ends of four of the Royal Welsh hanger scabbards, two of which retain their original iron hooks. They are constructed of tooled leather over a thin wooden liner. Of these, two were repaired after they arrived at Colonial Williamsburg with leather reinforcements around their hooks, the smaller of which is an incorrect replacement. The Colonial Williamsburg Foundation.

The tips of three of the Royal Welsh hanger scabbards, which still retain their sheet brass chapes, each embellished with two incised lines and two "V" shaped notches. The Colonial Williamsburg Foundation.

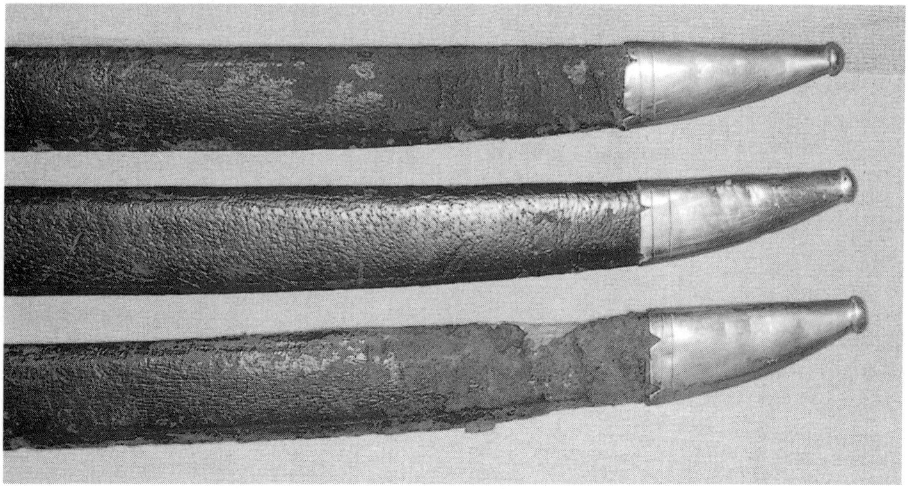

The Halberds

With the exception of the lone hanger and scabbard displayed over an archway, the halberds from Flixton Hall are the only martial weapons depicted in situ in surviving pictures of the mansion. A photograph taken between 1863 and 1871 shows four of the halberds mounted to the south wall of the great hall forming two "Xs." Two helmets are suspended above, with what is possibly a Royal Welsh infantry hanger hanging down from one set.

In a photo similar to the c. 1907 one with the sword over the arch, six of the seven halberds are shown in two upright racks of three, but it is unknown where the seventh was. These racks were in front of the two columns supporting the archway between the great hall and the grand staircase.

All seven pieces are identical, and are similar in form to the standard British Army serjeant's halberd of the 1750-1796 era, but they differ in minor details. That they differ from the common Ordnance pattern[78] may indicate that they were procured privately. All are hafted in darkened ash with plain iron ground points. The heads have a double-edged spear point with a median ridge and a crescent shaped blade opposite

Four of the halberds displayed on the south wall of the great hall at Flixton Hall, c. 1870 (a detail at top of page 35). The blurry item hanging down from the two at left may be one of the Royal Welsh hangers. Courtesy of Lady Darell.

a down curving beak point at the rear. They do not have some of the commonly encountered turned embellishments on the socket and are generally plainer than most serjeant's halberds of the period.

One piece is engraved "G" "G," which Neal erroneously speculated stands for the "Grenadier Guards," a name which the 1st Regt. of Foot Guards didn't acquire until 1816. It seems more likely that these two letters are the initials of a former owner or user. Since they are not of the Ordnance pattern, perhaps they were purchased for use by one of the Militia units to which the Adairs were agents.

Neal probably came up with the Grenadier Guards association because it was the regiment to which General Adair belonged. Neal also erroneously believed that many earlier members of the Adair family had been in this Guards regiment, but General Adair was the first and only in his direct line to have served in the Grenadier Guards.

Without clear regimental markings, it can not be determined if these weapons were used by serjeants of the Royal Welsh Fuziliers, who did carry such weapons during the period William Adair was their regimental agent. Since Adair had many other regiments, both regular army and militia, as clients during his career, it is possible that they could have come from almost any of them. One thing is for sure—during the Napoleonic Wars, Alexander Adair's 9th (South Elmham) Troop of Suffolk Yeomanry Cavalry had no use for them, or any of the other arms of the Royal Welsh Fuziliers.

A rack of three halberds against the northernmost arch support between Flixton Hall's grand staircase and great hall (a detail of photo on page 55). Courtesy of Lady Darell.

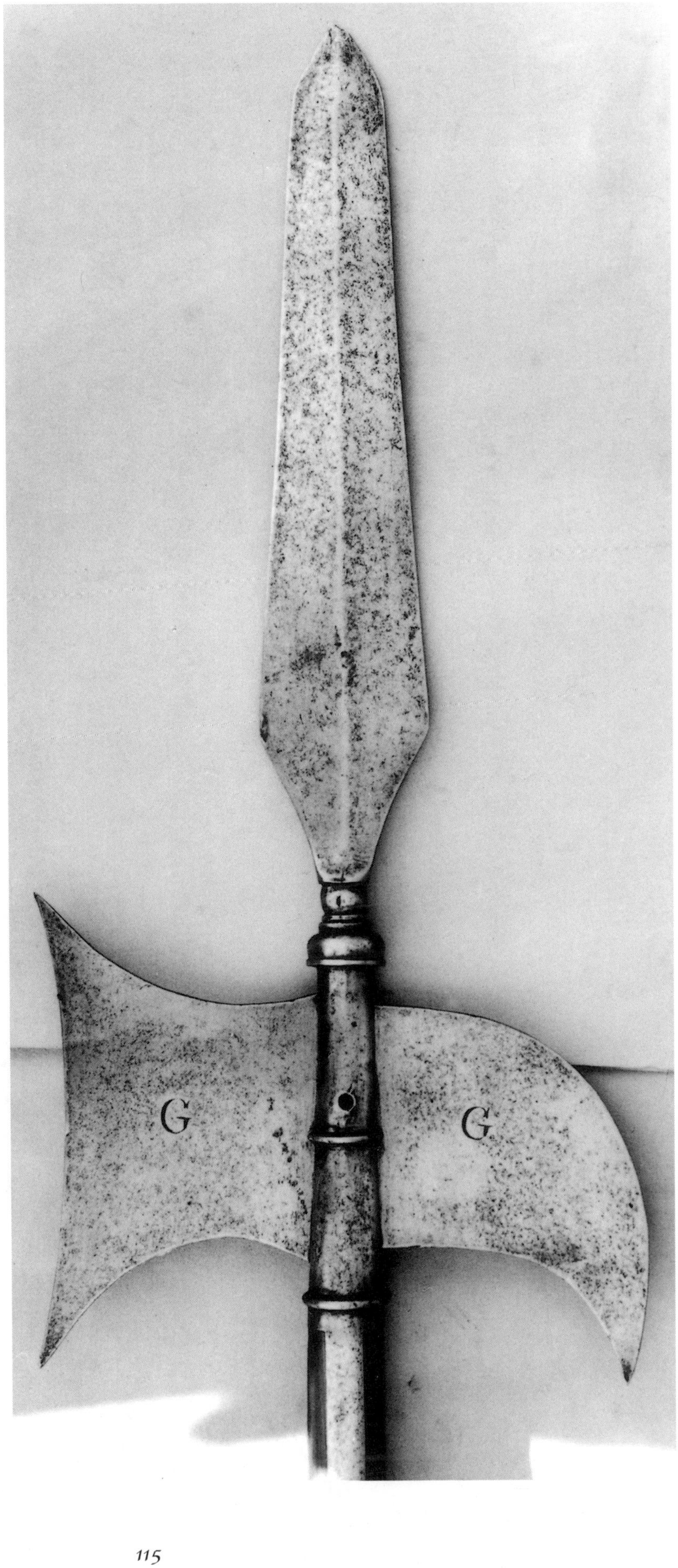

The lone "G G" marked halberd in a photograph sent by W. Keith Neal to Colonial Williamsburg with his letter of 2 November 1950. These marks are of unknown meaning, but certainly do not stand for "Grenadier Guards," as Neal speculated in his letter. The 1st. Foot Guards was not known as such until 1816, and neither William nor Alexander Adair had the agency of this Regiment. The hole through the bolster is for a screw, now missing, which helped secure the blade, removed to facilitate mounting. The Colonial Williamsburg Foundation.

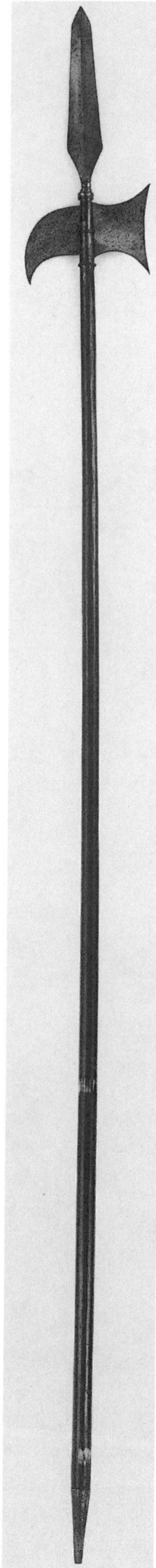

The iron butt cone of one of the halberds. The Colonial Williamsburg Foundation.

Overall view of one of the halberds. Total length 95 inches. The Colonial Williamsburg Foundation.

Interior of the Powder Magazine at Colonial Williamsburg, c. 1980-1990. Five of the halberds from Flixton Hall are seen in front, and immediately to the right, of the window. The Colonial Williamsburg Foundation.

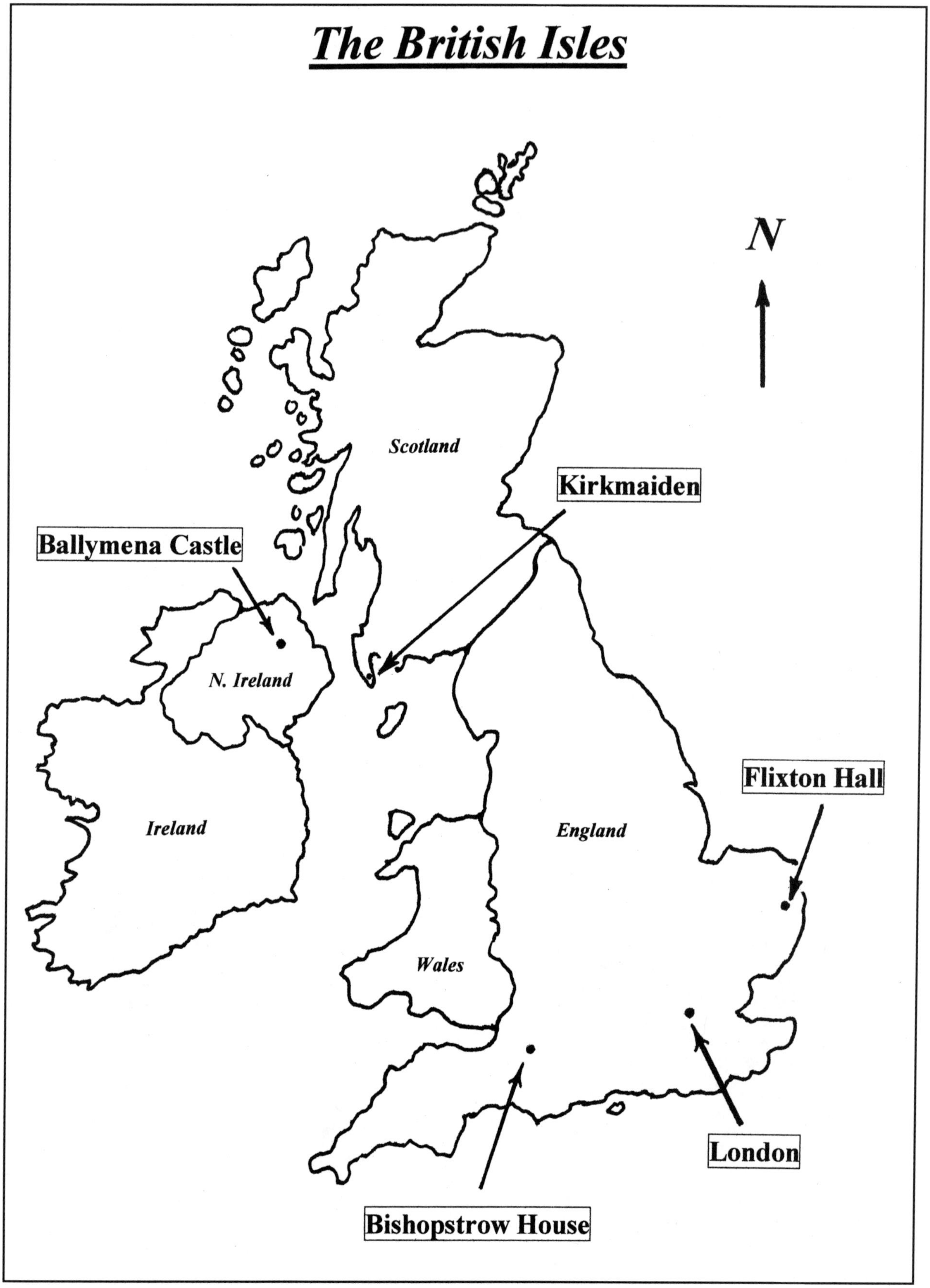

Map of Great Britain & Ireland denoting places discussed in the text.

Known Muskets from Flixton Hall

Pattern	Lock Marks	Weapon Number	Regimental Marks	Location	Accession #
-	1724/DC	-	-	PC, USA	-
1730	1727/Tower	G/No.62	RW	NAM	NAM 9002-55
1730	1728/Tower	B/No.9	RW	VFNHP	-
1730	1728/Es. Cole	Co.VIII/No.20	CDR	PC, USA	-
1730	1729/Cookes	No.19	RW	CWF	1951-335
1730	1730/Es. Cole	H/No.39	RWF	PC, USA	-
1730	1734/Jordan	C/No.2	RW	CWF	1951-338
1730	1734/Jordan	H/No.4	RWF	CWF	1954-84
1730	(replaced lock)	H/No.18	RW	RWFM	3026-A
1730/40	1741/Tower	H/No.16	RW	CWF	1951-325
1730/40	1742/Farmer	D/48	-	CWF	1951-331
1742	1742/Tower	CR/No.3	-	CWF	1951-337
1748	1747/Farmer	17/73	PW	CWF	1951-341
1748	1747/Farmer	19/71	PW	CWF	1951-339
1748	1747/Farmer	19/72	PW	CWF	1951-323
1748	1748/Wood	12/72	PW	CWF	1951-324
1748	1750/Farmer	13/73	PW	CWF	1951-329
1748	1750/Farmer	20/72	PW	CWF	1951-327
1748	(replaced lock)	20/71	PW	CWF	1951-322
1748	(replaced lock)	?	PW	unknown	-
1756	1756/Edge	4/73	PW	CWF	1951-326
1756	1756/Edge	5/72	PW	CWF	1951-340
1756	1756/Edge	7/71	PW	CWF	1951-332
1756	1756/Edge	8/73	PW	CWF	1951-334
1756	1756/Edge	10/72	PW	CWF	1951-336
1756	1756/Grice	5/73	PW	CWF	1954-94
1756	1756/Grice	10/71	PW	RA	XII-1580
1756	1756/Jordan	4/72	PW	CWF	1951-333
1756	1756/Perry	5/71	PW	CWF	1951-328
1756	1756/Galton	6/96	PW	CWF	1951-330
1756	1757/Edge	5/94	PW	PC, USA	-

Abbreviations

RW – ROYAL WELSH
RWF – ROYAL WELSH FUZILIERS
PW – Prince of Wales's device over "R-23" (the badge of the Regt.)
CDR – COLL. DALLWAYiS REGT. (39th.)
DC – Dublin Castle
CWF – The Colonial Williamsburg Foundation

NAM – National Army Museum, London
RA – Royal Armouries, Leeds
RWFM – The Royal Welch Fusiliers Museum, Caernarfon Castle
VFNHP – Valley Forge National Historic Park, The George C. Neumann Collection
PC – Private collection

Notes

1. There were a number of William Adairs in the family, some of whom were contemporaries. A William Adair is listed as the Chaplain of the 19th Regiment in 1767 (from January, 1764), at which time his relation of the same name was said Regiment's agent. This Chaplain William Adair may have been the one who served as Vicar of Flixton Church from 1754 until 1769.
2. Alan J. Guy, *Oeconomy and Discipline, Officership and Administration in the British Army, 1714-1763* (Manchester: University Press, 1985). This volume is an indispensable guide to understanding the management of the British Army of the period.
3. J. D. Turner, Army Agency, *The Journal for the Society for Army Historical Research*, Vol. 13, 1934: 28 & 29.
4. *Ibid.*, p. 28.
5. Guy, *Oeconomy and Discipline*, 59-62 & 143-6.
6. Alan J. Guy, *Minions of Fortune, The Regimental Agents of Early Georgian England, 1714-63* (Army Museum, 1985) 33 & 34.
7. *Ibid.*, 29-30.
8. The *Army List* for 1761 (London: J. Millan) lists "Mr. Bullock, Pall Mall" as the agent for the 58th Regiment. This could be either Thomas or his brother Henry, and is one of many appearances of a "Mr. Bullock" listed as an agent during this era.
9. Calcraft died in August 1772, leaving an estate valued at around 250,000 Pounds. When William Adair died in 1783, his estate was worth at least that much, in addition to his real estate holdings. See Alan Guy's "Minions of Fortune..." for more about John Calcraft.
10. "The Succession of Colonels to All His Majesties Land Forces from their Rise to 1746," p.1.
11. *Army List*, 1761.
12. Major E. L. Kirby, *Officers of the Royal Welch Fusiliers (23rd Regiment of Foot), 16 March 1689 to 4 August 1914* (Capel Curig: private publication, 1997) 71.
13. *Ibid.*
14. During the period, there was some latitude in the spelling of the Regiment's name. Both Welsh and Welch were common, as were Fusiliers, Fuziliers, Fuzileers and on occasion Fuciliers. In 1920 the regiment officially designated ROYAL WELCH FUSILIERS as the official spelling.
15. *The Gentleman's Magazine*, XXXI, January 1761: 22.
16. *Illustrated London News*, 19 December 1846: 393. Included in the report of the fire was a list of paintings known to have been at Flixton Hall, all of which were destroyed in the fire. Amongst these was one erroneously entitled "General Huss of Elling House," which is certainly General Huske of Ealing House.
17. There was apparently some intermarriage between the Adairs and the Keppels. There is a record of the birth of an Elizabeth Adair to Robert Adair and Caroline Keppel in Marylebone, London in 1769 (IGI).
18. The 1767 Army List records "Messrs. A. Adair and Bullock, Pall Mall" as the agents to the 19th Regiment of Foot.
19. It is not impossible that some of the Army Agency records of William and Alexander Adair were destroyed in the fire of 1846. Various late 1990s and early 2000s manuscript auctions in the U.S. have offered at least three documents that certainly came from the military papers of William and Alexander Adair (two were purchased by the author). This almost certainly points to a dispersal of such material at some point in the past by either the Adair family, or perhaps the Bullock family.
20. *Illustrated London News*, 26 December 1846: 404.
21. Rev. Alfred Suckling, *The History and Antiquities of the County of Suffolk*, Vol. I. (London: W. Hughes, 1846) 200.
22. There is other evidence that William Adair had Flixton Hall's moat and bridge removed shortly after his acquisition of estate. A map of the Flixton Estate, purported to be c. 1760, is in the Adair Papers at the Suffolk Record Office, Lowestoft (947/A1/1). Although the outline of Flixton Hall is not depicted accurately, an enclosure clearly surrounds the mansion and its gardens, and there is neither a moat, nor a bridge, depicted.
23. William Adair's will pertaining to his personal estate, proved 16 July 1783, is on file at the Public Records Office at Kew, London, and is referenced as PROB 11/1105.
24. A possible hint as to the identity of Jane (Adair) Brice's mother, or possibly a lover of William Adair's, appears in the will. A very generous bequest of a 5000 Pound 3% bank annuity was left to an apparently unmarried "Miss Charlotte Jones of Pall Mall," for "her own proper use and benefit." On just about all of the other bequests William Adair made, he stated his relationship to the beneficiary, but not in this case.
25. Information concerning the Brice family lineage and history came from an internet web page put up by Leslie Neeland, a student at Yale University as of late 2000 (see the bibliography for the site address).

    Edward and Jane Brice's eldest son wasn't Arthur, it was William Adair Brice, who unfortunately pre-deceased his grandfather and namesake by a little more than a year. He is listed by Neeland as "first Lieutenant of His Majesty's Ship Agamemnon and was slain in Rodney's action, 12 April 1782." This great naval battle, known as the Battle of the Saintes, saw the defeat of the Comte de Grasse off the coast of Dominica. Perhaps the death of his eldest grandson was relevant to the way in which William Adair divided up his estate?
26. Arthur Brice's service records came from period *Army Lists* and Col. D. MacKinnon, *Origin and Services of the Coldstream Guards*, Vol. II. (London: Richard Bentley, 1833). Thanks to Vincent J-R Kehoe for his help in finding this information.
27. William Adair was commissioned Lieutenant in the 25th Regt on 31 October 1776 and Captain on 1 June 1780.
28. It is possible that Robert Shafto was the subject of the famous 18th c. song "Bobby Shafto," now a popular children's song.
29. Although a kinship with the Shafto-Adair family has not been established, there was Cuthbert Shafto in the Royal Welsh Fusiliers during the time William Adair was its agent. Cuthbert obtained a 2nd Lieutenancy on 15 October 1753, and received the first commission issued using the term "Twenty Third Regiment of Foot" in addition to the name of the Regiment. On

26 December 1755 Shafto was promoted to Lieutenant, and left the Royal Welsh for the 75th Regiment on 24 November 1759.

30. *Bungay & Suffolk Herald*, 23 December 1846.
31. *Norwich Mercury*, 19 December 1846.
32. Portraits of the Duke of Cumberland, by Sir Joshua Reynolds, Sir Charles Saunders, by the same, Lord Anson, Admiral Keppel, General Hodgson, Lord Albemarle, Lord George Lennox, and General Robert Napier were amongst those destroyed by the fire (*Illustrated London News*, 19 Dec. 1846). William Adair had been the agent to all of the above listed men at one time or another.
33. Thanks to Glenn Earl for bringing these photographs to my attention.
34. Almost no documentary evidence pertaining to the various alterations to Flixton Hall has survived. As a result, Salvin has erroneously received the credit for the Hall's appearance during the post-1900 period, when most of the surviving photographs of it were taken. The extravaganza seen in these photographs is a result of the 1888-92 alterations by Wade, evidenced by the few known photographs taken after the fire of 1846 and before the later alterations.
35. Jill Allibone, Anthony Salvin, *Pioneer of Gothic Revival Architecture* (Columbia: University of Missouri Press, 1987) 172.
36. Thanks to Fred Mahnken of the 446th Bomb Group for his recollection of this event. Also see Ed Castens, *The 446th Revisited, The Story of Brogger's Bungay Buckaroos* (Santa Ana: 446th Bombardment Group Association, 1997) 55.
37. Sir Allan Shafto Adair, *A Guards General* (London: Hamish Hamilton, 1986) 188.
38. On 13 July 1948, the properties of the Flixton Hall Estate were offered at auction under the direction of General Adair by Messrs. Bidwell & Sons, of London, Cambridge and Ipswich.
39. Adair, 102 & 194. Perhaps if General Adair's only son and heir had not been killed in WWII, he may have decided to retain Flixton Hall and its contents to pass on to him.
40. *Eastern Daily Press*, 6 June 1952, and the *Beccles Journal*, of the same date.
41. That the roof leads were removed long before the demolition began is evidenced by a comparison of the 1944 aerial views with others taken in June of 1951 in Cambridge University's Collection of Aerial Photographs. The leads are plainly visible in the 1944 photos, and are missing in the 1951 views.
42. This paneling was installed at the Samson & Hercules before the demolition of Flixton Hall was complete. In addition to the paneling, two fireplaces from the Hall found their way to the ballroom too. These pieces from Flixton Hall were removed in the 1990s, and as of this writing, were being offered for sale on the Internet by a British firm which deals in architectural antiques.
43. A dovecote is a purpose-built structure intended to house doves and pigeons for the table of the main house. The one at Flixton Hall is in superb condition, and is now a protected building. It is built of the same red brick as the original parts of the main house.
44. There were a few lots of weapons listed in the sale catalog, which were either rejected by Neal, or unknown to him during his two visits to Flixton Hall. These lots were;
    769—African trophy of twenty various weapons, including spears, swords, knobkerries, etc.,
    998—Eight rapiers and swords, various, from "The Small Hall."
    999—Six swords and sabres, various, from "The Museum."
    1000—Six rapiers, swords and sabres, various, from "The Museum."
    1015—Seven swords and daggers, various, from "The Museum."
    Pair of old revolvers, a pistol and sundries, from "The Museum"
    Some of these non-Royal Welsh weapons are visible on display in the mansion in J. Arthur Rank's 1947 production of *The Secret Tunnel*, filmed at Flixton Hall.
45. Two cased sporting guns from the Adair cabinet appeared in Christie's 8 November 1995 sale of the William Keith Neal Collection. They were Lot 31, a James Wilkinson & Son wildfowling gun, c. 1850, which brought 2900 Pounds (hammer), and Lot 154, a Manton flintlock fowler made for Sir Robert Shafto Adair in 1820, which brought 9000 Pounds (hammer). However, Christie's 9 November 2000 sale of firearms from the W. Keith Neal Collection listed nothing from Flixton Hall.
46. The whereabouts of this lone spontoon are unknown, since it was not offered or sold to Colonial Williamsburg. Perhaps it was one of the spontoons serving as a flag staff in some of the early shots of the great hall (Figs. 27 & 50).
47. The condition in which these hangers were in when Neal bought them serves to illustrate that not only were they were issued (as opposed to most of the muskets), but also that they are found with the same defects seen on examples which did not go to Williamsburg.
48. Geoffrey Parnell, "The King's Guard Chamber A Vision of Power" *Apollo Magazine* (August 1994) 60-61. For more information concerning arms displays in English Royal residences and the Tower of London, consult the excellent articles written by this author listed in the bibliography.
49. *Ibid.*
50. Sarah Bevan, "Three-Dimensional Decorations, Displays of Arms in English Country Houses" *Country Life*, (October 1985) 1229-1230 for a discussion of the arms displays at Chevening House. I am also grateful to Dr. DeWitt Bailey for providing additional information on these arms displays.
51. *Illustrated London News*, 26 December 1846: 404.
52. Ivor Noel Hume, *A Guide to the Artifacts of Colonial America* (New York: Vintage Books, 1991) 254.
53. A photograph taken of the Great Hall and grand staircase c. 1870 show the halberds displayed on the south wall of the Great Hall. By the turn of the century, these halberds were displayed in racks in front of the arches facing the grand staircase. Descriptions of Flixton Hall from as early as the 1820s mention "armour" displayed in the Great Hall, but it appears that as alterations took place, the weapons displays were moved about.
54. The author firmly believes that William Adair formed the collection of arms at Flixton Hall. However, it is not impossible that these weapons, or some of them, were once the property of General Huske, and were inherited by William Adair in 1761 as part of the residue of Huske's estate.
55. Guy, *Oeconomy and Discipline*, 171.
56. Howard L. Blackmore, *Gunmakers of London, 1350-1850* Vol. I (York: George Shumway, 1986) 204.
57. George Moller, *American Military Shoulder Arms* Vol. I (Niwot: University Press of Colorado, 1993) 54-56, and George C. Neumann, *Battle Weapons of the American Revolution* (Texarkana: Scurlock Publishing, 1998) 56.

58. Neumann, *Battle Weapons*, 57.
59. In the c. 1915 pamphlet "Records of Flixton," Rev. A. A. Toms mentions that the weapons of Alexander Adair's Suffolk Yeomanry unit were "preserved in the armoury at the hall." While these were certainly the group of arms covered by this study, Flixton Hall did not have an armoury proper. Toms apparently noted the weapons displayed about the Great Hall, and applied the family legend as concerns their origin. Brown's 1957 article for *East Anglian Magazine*, mentions that the arms of the Yeomanry unit "were preserved in the armoury at Flixton Hall until its contents were sold in 1950," likely based on Toms' erroneous attribution.
60. What is believed to be the lone surviving musket of those provided by Freeman is now at the Royal Armouries (XII.1749), and is illustrated as Figs. 62 & 63 It was formerly in the museum of the Royal United Services Institution, and appears on p. 163 as item #2389 in their 1908 Catalog. Blackmore briefly discussed it, and its markings, on pp. 44, 282 & 283 and illustrated the top of the buttstock on p. 76 of British Military Firearms.... See DeWitt Bailey's listing under Freeman, James, on p. 44 of his *British Board of Ordnance Small Arms Contractors 1689-1840* (Rhyl: W. S. Curtis Publishers, 1999) for details concerning the issuance of this musket.
61. For a detailed description of the Pattern 1730 musket, see DeWitt Bailey, *Pattern Dates for British Ordnance Small Arms 1717-1783* (Gettysburg: Thomas Publications, 1997) 9-12, 23-25, 31, 49-50, 55-56 & 64.
62. See pp. (insert page # here) for information concerning "contingent" men and monies. Also see Guy, *Oeconomy and Discipline*, 127 & 128.
63. Thanks to Mr. David Bownes, Assistant Curator at The Royal Welch Fusiliers Regimental Museum at Caernarfon Castle, for providing details on this musket. This musket was discussed and illustrated by it previous owner, Ted Egly, in the January 1970 issue of *The Brigade Dispatch*, Vol. VII No. 1: 5-6. This magazine is the publication of the Brigade of the American Revolution, a North American organization dedicated to the re-enactment and interpretation of said period.
64. By the time a lock made in 1748 came into stores, the Board of Ordnance was no longer setting up Pattern 1730 muskets, and hadn't been for years.
65. This work would have been paid for by the respective Captains, out of "contingent" monies.
66. W. K. Neal & D. H. L. Back, Great British Gunmakers, 1540-1740 (Norwich: Historical Firearms, 1984) 443. The photo of the sideplate at the bottom of the page is of a different musket.
67. These ramrods have light pitting on the areas exposed to the air, which clearly delineates the areas protected by the ramrod pipes. When ramrods were matched to their respective muskets, it was apparent that these pitting patterns did not line up in many cases. This indicates that some of the ramrods spent many years in muskets other than those to which they were numbered.
68. J. A. Houlding, *Fit for Service, The Training of the British Army, 1715-1795* (Oxford: Clarendon Press, 1981) 141, note 84.
69. It is also possible that Pattern 1756 muskets were not yet in store when the Second Battalion Royal Welsh Fuziliers were armed.
70. For a detailed study of the Pattern 1727 bayonet, see Erik Goldstein, *The Socket Bayonet in the British Army, 1687-1783* (Lincoln: Andrew Mowbray, 2000) 49-54.
71. Guy, *Oeconomy and Discipline*, 59 & 147.
72. The worn, and in some cases damaged, condition of the hangers confirms this. Their overall condition is very consistent, more likely the result of use in the field, not from subsequent abuse.
73. Four of the weapon lots in the 1950 sale of Flixton Hall's contents contained a combined total of about 2 dozen swords (exclusive of the African ones), all listed as contents of "The Museum." Since this room was on the second floor above the Billiards Room, perhaps Neal had not known about the weapons kept there. Had he known about them, he would likely have bought them too, unless they were inconsequential.

    Regardless, it is not impossible that some of these swords were Royal Welsh Fuzilier hangers, which Neal did not purchase for one reason or another.
74. The bimetallic composition of many sword and bayonet blades of this period was intended to give the benefits of both a hard steel cutting edge and a flexible iron base.
75. A. D. L. Cary and Stouppe McCance, *Regimental Records of the Royal Welch Fusiliers* Vol. I (London: Royal United Services Institution, 1921) 122, 124 & 125.
76. *Ibid.*, p. 124.
77. The 16 February 1757 return indicates that "complete" strength for the Royal Welsh was between 700 and 730 enlisted men per battalion, formed in 10 companies of 70-73 men. This was for both the First Battalion, and those companies of the Second that had been formed by that date.
78. For a common Ordnance-pattern halberd of the mid-18th century, see George Neumann, *Swords and Blade of the American Revolution* (Harrisburg: Stackpole Books, 1973) 195, # 7.PA.

Bibliography

Periodicals & Newspapers

Beccles Journal, 6 June 1952.

Bungay & Suffolk Herald, 23 December 1846.

East Anglian Daily Times, 2 February 1935 and 19 November 1953.

East Anglian Magazine, February 1948
March 1957
June 1957
July 1959.

Eastern Daily Press, 23 July 1934
18 September 1951
6 June 1952
25 September 1952
8 August 1953
24 August 1953
30 July 1990

Eastern Evening News, 19 November 1953.

The Gentleman's Magazine, XXXI, London, January,1761.

Illustrated London News, 19 December 1846 and 26 December 1846.

Norwich Mercury, 19 December 1846, 26 December 1846 and 16 January 1847

The Star, 9 June 1952.

Books

Adair, Sir Allan Shafto, A Guards General, London: Hamish Hamilton, 1986.

Ahearn, William, "British Longarms at Lexington," Man at Arms, Vol. 20 No. 2 (April 1998): 15-24.

Allibone, Jill, Anthony Salvin, Pioneer of Gothic Revival Architecture, Columbia: University of Missouri Press, 1987.

Angus, W., The Seats of the Nobility and Gentry in Great Britain and Wales in a Collection of Select Views, Islington, 1787.

Bailey, DeWitt, Pattern Dates for British Ordnance Small Arms, 1717-1783, Gettysburg: Thomas Publications, 1997.

British Board of Ordnance Small Arms Contractors 1689-1840, Rhyl: W.S. Curtis Publisher, Ltd., 1999.

British Military Longarms, 1715-1865, London: Arms and Armour Press, 1986.

"British Military Small Arms," Journal of the American Society of Arms Collectors, 71 (October 1994): 2-14.

"Development of the British Ordnance Musket Lock, Part I," International Collector Magazine, Vol. 1 No. 1 (1995): 47-50.

Barker, Horace R., East Suffolk Illustrated, Bury St. Edmunds: F.G. Pawsey & Co., 1909.

Bedford, Clay P., And Stephen V. Grancsay, Early Firearms of Great Britain and Ireland, Greenwich: New York Graphic Society (for the Metropolitan Museum of Art), 1971.

Bevan, Sarah, "Three-Dimensional Decorations, Displays of Arms in English Country Houses," Country Life, (1985): 1229-1238.

Blackmore, Howard, British Military Firearms, 1650-1850, London: Herbert Jenkins, 1961.

Gunmakers of London, 1350-1850, York: George Shumway Publisher, 1986.

Bidwell & Sons, Messrs., The Flixton Estate (sale of by auction), Cambridge, Ipswich & London: Bidwell & Sons, 1948.

Brown, Archibald C., The Adair Family of Flixton Hall, East Anglian Magazine (March 1957): 254-260.

Burke's Peerage and Baronetage, London, Burke's Peerage, Ltd., 1939 & 1968.

Carroll, Donald, "The Brown Bess Musket Regimentally Marked," Gun Report, Vol. 35 No. 8 (January 1990): 36-42.

Cary, A.D.L., and Stouppe McCance, Regimental Records of the Royal Welch Fusiliers, Vol. I, London: Royal United Services Institution, 1921.

Castens, Ed, The 446th Revisited, The Story of Brogger's Bungay Buckaroos, Santa Ana: 446th Bombardment Group Association, 1997.

Copinger, W.A., The Manors of Suffolk, Manchester: Taylor, Garnett, Evans & Co., Ltd., 1911.

Cromwell, Thomas Kitson, Excursions in the County of Suffolk, London, Longman & Co., 1818.

Darling, Anthony D., Redcoat and Brown Bess, Bloomfield: Museum Restoration Service, 1970.

"Mid 18th Century British Military Swords with Open "S" Paneled Guards," American Society of Arms Collectors, 49 (October 1983): 2-9.

"Mid 18th Century British Military Swords with Open "S" Paneled Guards, an Update," American Society of Arms Collectors, 62 (May 1990):2-8.

Davis, W.J., The Nineteenth Century Token Coinage of Great Britain, Ireland, The Channel Islands and the Isle of Man, London: B.A. Seaby Ltd., 1969.

Davy, David Elisha, A Journal of Excursions Through the County of Suffolk, 1823-1844, Woodbridge: Boydell Press for the Suffolk Records Society (vol. XXIV), 1982.

Davy, Henry, A Series of Etchings Illustrative of the Architectural Antiquities of Suffolk Accompanied with an Historical Index, Southwold, 1827.

Davy, Henry, Views of the Seats of the Noblemen and Gentlemen in Suffolk, Southwold, 1827.

Egly, Ted, "A Musket of Captain William Hickman, British 23rd Regiment. ñ The Royal Welsh Fusiliers" The Brigade Dispatch, Vol.VII, No.1 (January 1970): 5-6.

Glover, Michael, That Astonishing Infantry, The History of the Royal Welch Fusiliers 1689-1989, London: Leo Cooper, 1989.

Goldstein, Erik, The Socket Bayonet in the British Army, 1687-1783, Lincoln: Andrew Mowbray, 2000.

"Infantry Hangers of the Royal Welch Fusiliers, 1742-1784," Man at Arms, Vol.20, No.6 (December 1998): 18-23.

Guy, Alan J., Oeconomy and Discipline, Officership and Administration in the British Army, 1714-1763, Manchester: University Press, 1985.

Colonel Samuel Bagshawe and the Army of George II, 1731-1762, London: Army Records Society, 1990.

"Minions of Fortune, The Regimental Agents of Early Georgian England, 1714-63," Army Museum '85 (The National Army Museum, 1985): 31-42.

Hawkins, Edward, Medallic Illustrations of the History of Great Britain and Ireland to the Death of George II, Vol. II, London: Spink & Son, 1978 (originally published in 1885).

Hood, Graham, The Governor's Palace in Williamsburg, Williamsburg: The Colonial Williamsburg Foundation, 1991.

Houlding, J.A., Fit for Service, The Training of the British Army, 1715-1795, Oxford: Clarendon Press, 1981.

Hume, Ivor Noel, A Guide to the Artifacts of Colonial America, New York: Vintage Books, 1991.

Kenworthy-Brown, John, et al, Burke's and Savills Guide to Country Houses, Vol III, East Anglia, London, Burke's Peerage Ltd., 1981.

Kirby, Major E.L., Officers of the Royal Welch Fusiliers (23rd Regiment of Foot), 16 March 1689 to 4 August 1914, Capel Curig: private publication, 1997.

Leslie, N.B., The Succession of Colonels of the British Army from 1660 to the Present Day, London: Society for Army Historical Research (Special Publication #11), 1974.

Lindsay, Col. O.J.M., Major General Sir Allan Adair, Bt... Obituary, Guards Magazine, 1988, pp. 159-160.

MacKinnon, Col. D., Origin and Services of the Coldstream Guards, London: Richard Bentley, 1833.

Moller, George, American Military Shoulder Arms, Vol. I., Niwot: University Press of Colorado, 1993.

Moore, Andrew W., John Sell Cotman 1782-1842, Norwich: Norfolk Museums Service, 1982.

Neal, W. Keith, And D.H.L. Back, Great British Gunmakers, 1540-1740, Norwich: Historical Firearms, 1984.

Neale, J. P., Views of the Seats of Noblemen and Gentlemen in England, Wales, Scotland and Ireland, London: Sherwood Jones & Co., 1823.

Neeland, Leslie, The Brice Family Genealogy Homepage,
HYPERLINK http://www.mindspring.com/~dbn/Brice.html http://www.mindspring.com/~dbn/Brice.html, version of 14 August 2000.

Neumann, George C., Battle Weapons of the American Revolution, Texarkana: Scurlock Publishing, 1998.

Swords and Blades of the American Revolution, Harrisburg: Stackpole Books, 1973.

History of the Weapons of the American Revolution, New York: Harper & Row, 1967.

Neumann, George C., And Frank J. Kravic, The Collectors Illustrated Encyclopedia of the American Revolution, Harrisburg: Stackpole Books, 1975.

Parnell, Geoffrey, "The Early History of the Tower Armouries," Royal Armouries Yearbook, Vol. I (1996): 45-52.

"Arming the Tower, The Small Armoury at the Tower of London," Apollo Magazine, (February 1993): 82-85.

"The King's Guard Chamber, A Vision of Power," Apollo Magazine, (August 1994): 60-64.

Pevsner, Sir Nicolaus, The Buildings of England; Suffolk, Middlesex: Penguin Books, 1974.

Pine, John (engr.), A Representation of the Cloathing of His Majesty's Household and all of the Forces upon the Establishment of Great Britain & Ireland, By order of The Duke of Cumberland, London, 1742.

Priest, Graham, Brown Bess Bayonets, 1720-1860, Norwich: Tharston Press, 1986.

Sheppard, F.H.W. (ed.), Survey of London, Vol. XXX, London: London City Council, 1960.

Stanford, Broom & Stanford, Catalog of the Sale at Flixton Hall, Near Bungay, Suffolk, 27 November to 1 December 1950, Halesworth: Stanford, Broome & Stanford, 1950.

Strachan, Hew, British Military Uniforms, 1768-1796, London: Arms and Armour Press, 1975.

Strong, Sir Roy, The Destruction of the Country House, 1875-1975, London: Thames and Hudson, 1975.

The Succession of Colonels to All His Majesties Land Forces from their Rise to 1746, London: J. Millan, 1746.

Suckling, Rev. Alfred, The History and Antiquities of the County of Suffolk..., Vol. I, London: W. Hughes, 1846.

Sumner, The Rev. Percy, "Portrait of Lieutenant Joseph Sabine, Royal Welsh Fuziliers, 1740," The Journal of the Society for Army Historical Research, Vol. XXII No. 90 (Summer, 1944): 219-220.

Toms, Rev, A.A., Records of Flixton, Bungay: Bernard Clay & Sons, Ltd., 1915.

Turner, J.D., "Army Agency," The Journal of the Society for Army Historical Research, Vol. 13 (1934): 27-37.

INDEX

The author, under the ruined arches of Flixton Hall's west wing, 25 March 1999.

THOMAS PUBLICATIONS publishes books about the American Colonial era, the Revolutionary War, the Civil War, and other important topics. For a complete list of titles, please visit our web site:

http://thomaspublications.com

Or write to:

THOMAS PUBLICATIONS
P.O. Box 3031
Gettysburg, PA 17325